MONROE COLLEGE LIBRARY
3 7340 01067604

D0105748

1001 COMPUTER WORDS

You Need to Know

Other titles in the series:

1001 Legal Words You Need to Know

1001 Financial Words You Need to Know

1001 COMPUTER WORDS

You Need to Know

JERRY POURNELLE

Editor

ERIN McKEAN

Series Editor

OXFORD

UNIVERSITY PRESS

2004

Oxford University Press

Oxford New York
Auckland Bangkok Buenos Aires Cape Town Chennai
Dar es Salaam Delhi Hong Kong Istanbul Karachi Kolkata
Kuala Lumpur Madrid Melbourne Mexico City Mumbai
Nairobi São Paulo Shanghai Taipei Tokyo Toronto

Published by Oxford University Press, Inc.
198 Madison Avenue, New York, New York, 10016
http://www.oup.com/us

Library of Congress Cataloging-in-Publication Data

1001 computer words you need to know / edited by Jerry Pournelle.
p. cm.
Includes bibliographical references.
ISBN 0-19-516775-9 (hardcover : alk. paper)
1. Computers. 2. Computers--Dictionaries. I. Title: One thousand and one computer words you need to know. II. Pournelle, Jerry, 1933–
QA76.5.A125 2004
004'.03--dc22

2004003035

Printing number: 9 8 7 6 5 4 3 2 1

Printed in the United States of America
on acid-free paper

CONTENTS

Preface vii

Editorial Staff viii

Introduction ix

Using This Dictionary xiii

Pronunciation Key xv

1001 COMPUTER WORDS YOU
NEED TO KNOW 1

Which Operating System Is Right For You? 16

Basic Troubleshooting 55

The Ten Worst E-mail Mistakes 76

Buying a Computer 88

Dealing with Spam 108

How to Protect Yourself from Hoaxes,
Frauds, and Identity Theft 130

How to Shop Safely Online 143

Eight Simple Rules for Kid-friendly Computing . . 171

The Ten Best Tools and Peripherals You
Didn't Know About 196

Multimedia and You 210

Ten Great Computer Games You Should (or Maybe Shouldn't) Know 219

Do You Need a PDA? 224

Fifty Web Sites You Should Know 227

A Quick Guide to Writing Online English 235

PREFACE

This book, *1001 Computer Words You Need to Know*, is part of an Oxford University Press series of concise, helpful guides to the vocabularies of significant fields. We have distilled and enhanced the general dictionary entries (taken from our groundbreaking *New Oxford American Dictionary*) to make useful, browsable books that present the most important words needed to understand a particular topic, selected and updated by recognized experts. By stripping away the words you don't immediately need, we bring these complicated topics into sharper focus.

In addition to these essential 1001 words, we've added explanatory essays about related topics, both serious and lighthearted, and a list of essential web sites for further browsing, making this more than just a dictionary—a truly practical and entertaining all-around guide and reference book.

Erin McKean
Senior Editor, U.S. Dictionaries,
Oxford University Press

EDITORIAL STAFF

Erin McKean, *Series Editor*
Constance Baboukis, *Managing Editor*
Carol-June Cassidy, *Project Editor*
Christine Lindberg, Grant Barrett, Enid Pearsons,
Evan Schnittman, Dylan Wilbanks, *Editors*
Ryan Sullivan, *Editorial Assistant*

INTRODUCTION

I wrote my first books on a typewriter. I suppose most of you know what those are: odd devices that put letters and words on paper without first encoding them as bits and storing them in memory, on fast spinning metal disks, or on tape. Typewriters have this problem: the text is on paper, and when you edit the text, generally with a red pencil, you have to pay someone to re-type the whole thing, or retype it yourself. The first method is expensive. The second is exquisitely boring.

Thus when, back in the late 70's, I discovered small computers and "word processing" programs it was cause for joy. To my wife's horror I borrowed enough money to get a small computer and a Diablo electric typewriter that would make a printed manuscript from electronic files. If you want to see what all that looked like, my old computer, Ezekiel, is on display in the Hall of Communications and Computers in the Smithsonian Museum of American History in Washington, DC. And that ought to be sufficient to establish my credentials for writing an introduction to this book.

I joined the computer revolution joyfully, because while I don't hate writing, I do hate retyping. When I first started working with computers, I would write a draft of a book or a magazine article on the computer screen and save it to disk (which meant floppy disk in those days). I would then print it out, edit on paper, and then manually type the edited changes into the electronic version of the piece. Then I'd let the computer print out another draft, edit that, enter the changes, etc., until I couldn't find anything else to change. I hated to retype things, but old Ezekiel didn't mind at all.

Later I found that was needless work, and now most of my work is done on the computer and submitted electronically, so there is never a paper copy at all. Editing is done on screen, and in fact I seldom read anything I have written without making a few changes here and there.

This all sounds simple, and now it generally is, but it wasn't always simple. Even today there can be technical complications if you don't know what you're doing. It was worse in the early days. Back before the World Wide Web and ubiquitous e-mail, getting a book or an article sent over the wire was very hard to do, and involved working with computer experts. Sometimes it involved language problems as well: in 1983 I had to file a story from Liechtenstein, and I guarantee you that using that Principality's telephone system (which was part of their Post Office) was hideously complex. Actually, I never was able to do it: I printed my 8,000-word column on a tiny little Radio Shack thermal printer which produced a tape two inches wide and many yards long. The tape was sent to the *Byte* editorial offices in Peterborough, New Hampshire by Emory express delivery, and a tech editor had to type all 8,000 words into a word processor. I never was able to look that editor in the eye again. In general, though, I managed to solve most of my technical problems over the years.

Along the way I developed one of Pournelle's Laws: if you don't know what you're doing, work with people who do. In my case this meant working with people who understood computers and computer technology.

And that is an art all by itself.

People who use computers fall into two categories: wizards and the rest of us. The goal of the computer industry executives is to make these machines "user friendly," usable by nearly everyone. The goal of the wizards is to make it impossible to use the machines without hiring wizards as consultants. Since the executives—the suits—have no choice but to employ the wizards in their efforts to make the machines user friendly, you can see they start with a considerable handicap.

It's not so much that the wizards aren't trying to make things easier; it's that often they can't. They can't explain even comparatively simple operations to outsiders, because the explanation requires them to use technical terms, words that few outside the computer profession understand; so while the operation may look simple, and for the user it may be simple, understanding what's going on is another matter, and it's made more complex by the words the wizards are compelled to use.

Every profession develops a jargon. Automobile mechanics have a jargon, and some of them employ it to intimidate customers. The medical profession has a jargon: when a physician tells you that you have *lumbago,* it sounds much more learned than if he says you have a pain in your lower back. The interesting thing is that *lumbago* means exactly that: a lower backache, which is after all what you just told the doctor. You knew you had a lower backache, even if you didn't know you had lumbago.

Talking to computer experts can produce the same result, only worse. Computer wizards haven't had as long to develop a jargon as the physicians have had, but it's just about as rich, far richer than the jargon of automobile mechanics or plumbers—and it's entirely incomprehensible to most of us.

If you want to know more about the lives of wizards, you can see them at work and play at www.userfriendly.org. You can also see what happens when ordinary users ask for tech support from wizards. Both parties get frustrated. Sometimes there's violence. And look at the cartoon for February 16, 1999 to get an idea of what can happen when a marketing guy—a suit—tries to enter the world of the beards, AKA wizards. It's funny but it's not pretty.

Fortunately, there's help in this book. Studying a book of computer words—learning the jargon—won't make you a wizard. (It may let you fool others into thinking you're a wizard, which is another story.) Most of you don't really want to become wizards. You just want to be able to use your machines, and when things go wrong, get some help from people who know what they're doing.

And for that there's no substitute for knowing the vocabulary.

You can't ask for help if you can't describe your problem; and you can't get help if you can't understand what the wizard is telling you, and believe me, I have seen both parts of that situation literally thousands of times in the thirty years or so I have been writing about computers.

The way out of those boxes is to learn the jargon. I won't pretend that it's always easy to do that, but some ways are more painful than others. Meanwhile, now that you have this book, you can look up things you don't understand, and find that often they make more sense than you thought. You just need the words.

In addition to the words themselves, there are explanatory sidebars scattered through the book. Use them to get a feel for what the bare words mean. When you're done you won't be a wizard, but you will understand more about this world that has crept up on us and now has taken over. And that is no small thing.

Jerry Pournelle
March, 2004

USING THIS DICTIONARY

The "entry map" below explains the different parts of an entry.

Syllabification

Pronunciation set off with slashes / /

ac•cess /'ækses/ ▸ **n.** the action or process of obtaining or retrieving information stored in a computer's memory: *this prevents unauthorized access or inadvertent deletion of the file.* —— Examples in *italic*

Subsenses signalled by ■

■ the opportunity to use a computer, files, data, etc.: *unauthorized user access.* ■ a way to connect to the Internet: *broadband access.*
▸ **v.** [trans.] (usu. **be accessed**) obtain, examine, or retrieve (data or a file). Grammar information in square brackets [ ]

USAGE: The verb **access** is standard and common in computing and related terminology, but the word is primarily a noun. Outside computing contexts, its use as a verb in the sense of 'approach or enter a place' is often regarded as nonstandard.

Usage notes provide extra information to help you understand the use or importance of the term.

sleep /slēp/ ▸ **n.** turn off (devices attached to a computer) to reduce power use; hibernate or place on standby.

Phrases section, phrases in **bold face**

PHRASES **put something to sleep** put a computer on standby while it is not being used.

Etymology section

ORIGIN Old English *slēp*, *slǣp* (noun), *slēpan*, *slǣpan* (verb), of Germanic origin; related to Dutch *slapen* and German *schlafen*.

vir•tu•al /ˈvərCHo͞oəl/ ▸ **adj.** not physically existing, but made by software to appear to do so: *a virtual computer.* See also **VIRTUAL REALITY**.

Derivative section, derivatives in **bold face**

DERIVATIVES **vir•tu•al•i•ty** /ˌvərCHo͞oˈælit̲ē/ **n.**

Cross references in **BOLD SMALL CAPITALS**

Main entries and other boldface forms

Main entries appear in boldface type, as do inflected forms, idioms and phrases, and derivatives. The words PHRASES and DERIVATIVES introduce those elements. Main entries and derivatives of two or more syllables show syllabification with centered dots.

Parts of speech

Each new part of speech is introduced by a small right-facing arrow.

Senses and subsenses

The main sense of each word follows the part of speech and any grammatical information (e.g., [intrans.] before a verb definition). If there are two or more main senses for a word, these are numbered in boldface. Closely related subsenses of each main sense are introduced by a solid black box. In the entry for **access** above, the main sense of "the action or process of obtaining or retrieving information stored in a computer's memory" is followed by a related sense, "the opportunity to use a computer, files, data, etc."

Example sentences

Example sentences are shown in italic typeface; certain common expressions appear in bold italic typeface within examples, as in the entry for **box** above: *broadband access.*

Cross references

Cross references to main entries appear in small capitals. For example, in the entry **virtual** seen previously, a cross reference is given in bold small capitals to the entry for **VIRTUAL REALITY**.

A Note About URLs in This Book

All URLS given in this book should be typed exactly as they appear. Any punctuation necessary for the sentence that is not part of the URL will be in parentheses, such as http://www.safernetworking.org(.) All of the URLs or web addresses given work at the time of publication. If you enter one that cannot be found by your browser, first check your typing! Many of these URLs are difficult to type correctly. If you still get an error message, you can try to enter the URL at http://www.archive.org/(,) a site that periodically archives web pages.

PRONUNCIATION KEY

This dictionary uses a simple respelling system to show how entries are pronounced, using the following symbols:

æ *as in* **hat** /hæt/, **fashion** /ˈfæsHən/, **carry** /ˈkærē/
ā *as in* **day** /dā/, **rate** /rāt/, **maid** /mād/, **prey** /prā/
ä *as in* **lot** /lät/, **father** /ˈfäTHər/, **barnyard** /ˈbärnˌyärd/
b *as in* **big** /big/
CH *as in* **church** /CHərCH/, **picture** /ˈpikCHər/
d *as in* **dog** /dôg/, **bed** /bed/
e *as in* **men** /men/, **bet** /bet/, **ferry** /ˈferē/
ē *as in* **feet** /fēt/, **receive** /riˈsēv/
er *as in* **air** /er/, **care** /ker/
ə *as in* **about** /əˈbowt/, **soda** /ˈsōdə/, **mother** /ˈməTHər/, **person** /ˈpərsən/
f *as in* **free** /frē/, **graph** /græf/, **tough** /təf/
g *as in* **get** /get/, **exist** /igˈzist/, **egg** /eg/
h *as in* **her** /hər/, **behave** /biˈhāv/
i *as in* **guild** /gild/, **women** /ˈwimin/
ī *as in* **time** /tīm/, **fight** /fīt/, **guide** /gīd/, **hire** /hīr/
ir *as in* **ear** /ir/, **beer** /bir/, **pierce** /pirs/
j *as in* **judge** /jəj/, **carriage** /ˈkærij/
k *as in* **kettle** /ˈketl/, **cut** /kət/
l *as in* **lap** /læp/, **cellar** /ˈselər/, **cradle** /ˈkrādl/
m *as in* **main** /mān/, **dam** /dæm/
n *as in* **honor** /ˈänər/, **maiden** /ˈmādn/
NG *as in* **sing** /siNG/, **anger** /ˈæNGgər/
ō *as in* **go** /gō/, **promote** /prəˈmōt/
ô *as in* **law** /lô/, **thought** /THôt/, **lore** / lôr/
oi *as in* **boy** /boi/, **noisy** /ˈnoizē/
o͝o *as in* **wood** /wo͝od/, **football** /ˈfo͝otˌbôl/, **sure** /SHo͝or/
o͞o *as in* **food** /fo͞od/, **music** /ˈmyo͞ozik/
ow *as in* **mouse** /mows/, **coward** /ˈkowərd/

p *as in* **put** /po͝ot/, **cap** /kæp/
r *as in* **run** /rən/, **fur** /fər/, **spirit** /'spirit/
s *as in* **sit** /sit/, **lesson** /'lesən/
SH *as in* **shut** /SHət/, **social** /'sōSHəl/, **action** /'ækSHən/
t *as in* **top** /täp/, **seat** /sēt/
t̲ *as in* **butter** /'bət̲ər/, **forty** /'fôrt̲ē/, **bottle** /'bät̲l/
TH *as in* **thin** /THin/, **truth** /tro͞oTH/
T̲H̲ *as in* **then** /T̲H̲en/, **father** /'fäT̲H̲ər/
v *as in* **never** /'nevər/, **very** /'verē/
w *as in* **wait** /wāt/, **quick** /kwik/
(h)w *as in* **when** /(h)wen/, **which** /(h)wiCH/
y *as in* **yet** /yet/, **accuse** /ə'kyo͞oz/
z *as in* **zipper** /'zipər/, **musician** /myo͞o'ziSHən/
ZH *as in* **measure** /'meZHər/, **vision** /'viZHən/

Foreign Sounds
KH *as in* **Bach** /bäKH/
N *as in* **en route** /äN 'ro͞ot/, **Rodin** /rō'dæN/
œ *as in* **hors d'oeuvre** /ôr 'dœvrə/, **Goethe** /'gœtə/
Y *as in* **Lully** /lY'lē/, **Utrecht** /'Y,treKHt/

Stress marks

Stress marks are placed before the affected syllable. The primary stress mark is a short vertical line above the letters ['] and signifies greater pronunciation emphasis should be placed on that syllable. The secondary stress mark is a short vertical line below the letters [,] and signifies a weaker pronunciation emphasis.

A

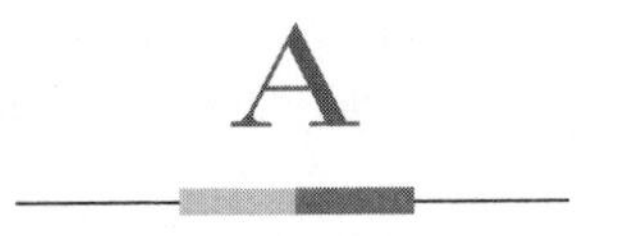

ac•cel•er•at•ed graph•ics port /æk'selə,rātid 'græfiks ,pôrt/ ▸ **n.** see **AGP**.

ac•cel•er•a•tor board /æk'selə,rātər ,bôrd/ (also **ac•cel•er•a•tor card**) ▸ **n.** an accessory circuit board that can be plugged into a desktop computer to increase the speed of its processor or input/output operations.

ac•cess /'ækses/ ▸ **n.** the action or process of obtaining or retrieving information stored in a computer's memory: *this prevents unauthorized access or inadvertent deletion of the file.*

■ the opportunity to use a computer, files, data, etc.: *unauthorized user access.* ■ a way to connect to the Internet: *broadband access.*

▸ **v.** [trans.] (usu. **be accessed**) obtain, examine, or retrieve (data or a file).

> USAGE: The verb **access** is standard and common in computing and related terminology, but the word is primarily a noun. Outside computing contexts, its use as a verb in the sense of 'approach or enter a place' is often regarded as nonstandard.

ac•cess con•trol list /'ækses kən,trōl ,list/ (abbr.: **ACL**) ▸ **n.** a function that manages the level of access to a computer directory or file each user of the system has (e.g., view a file, modify data, or execute a program) as set by a system administrator.

ac•cess pro•vid•er /'ækses prə,vīdər/ ▸ **n.** another term for **SERVICE PROVIDER**.

ac•cess time /'ækses ,tīm/ ▸ **n.** the time taken to retrieve data from storage.

ac•cu•mu•la•tor /ə'kyo͞omyəˌlāt̲ər/ ▸ **n.** a register used to contain the results of an arithmetical or logical operation.

ac•tive ma•trix /'æktiv 'mātriks/ ▸ **n.** an LCD display system in which each pixel is individually controlled. Also see **TFT**.

A•da /'ādə/ ▸ **n.** a high-level computer programming language used esp. in real-time computerized control systems, e.g., for aircraft navigation.

ORIGIN 1980s: from the name of *Ada* Lovelace, the daughter of Lord Byron. She assisted Charles Babbage in developing the first mechanical computer.

ad•bot /'ædˌbät/ ▸ **n.** a computer program that caches advertising on personal computers from an Internet-connected server and then displays the advertising when certain linked programs are being used: *click on the startup tab to view all the things that get loaded when you start Windows, and then uncheck anything that looks like the adbot software.*

add-in /'æd ˌin/ ▸ **n.** another term for **ADD-ON**.

add-on /'æd ˌän/ (also **add-in**) ▸ n. a printed circuit board, designed to provide an enhancement such as advanced audio or video, that can be inserted into a slot in a computer.

ad•dress /ə'dres; 'ædres/ ▸ **n.** a binary number that identifies a particular location in a data storage system or computer memory.

■ an e-mail address. ■ an IP address.

ad•dress•a•ble /ə'dresəbəl/ ▸ **adj.** relating to or denoting a memory unit in which all locations can be separately accessed by a particular program.

ad•dy /'ædē/ ▸ **n.** (pl. **ad•dies**) informal an address, especially an e-mail address: *I just sent a note to you and Jemily from my other addy.*

ADP ▸ **abbr.** automatic data processing.

ADSL ▸ **abbr.** asymmetric (or asynchronous) digital subscriber line, a form of DSL in which data is transmitted downstream, to the user, at a faster rate then it is transmitted upstream, from the user. ADSL is the most commonly used form of DSL available to home users.

ad•ver•game /'advərˌgām/ ▸ **n.** a downloadable or Internet-based computer game that advertises a brand-name product by featuring it as part of the game: *born of desperation and ingenuity, advergames,* *as they are called by marketers, are emerging at a time when Web surfers largely ignore more conventional forms of advertising.*

DERIVATIVES **ad•ver•gam•ing n.**

ORIGIN blend of *advertisement* and *game.*

a•gent /'ājənt/ ▸ **n.** an independently operating computer program, typically one set up to locate specific information on the Internet and deliver it on a regular basis: *in the future, there will be almost as few humans browsing the Net as there are people using libraries today. Agents will be doing that for most of us.*

ag•gre•ga•tor /'ægri,gātər/ ▸ **n.** an Internet company that collects information about other companies' products and services and distributes it through a single Web site.

AGP ▸ **abbr.** accelerated graphics port, a video graphics technology providing a connection to computer memory that allows 3D graphics to display easily on a computer monitor.

AI ▸ **abbr.** artificial intelligence.

Al•gol /'ælgôl/ ▸ **n.** one of the early high-level computer programming languages that was devised to carry out scientific calculations.

ORIGIN 1950s: from *algo(rithmic)* + the initial letter of **LANGUAGE**.

a•li•as /'ālēəs/ ▸ **n.** an alternative name or label that refers to a file, command, address, or other item, and can be used to locate or access it.

ORIGIN late Middle English: from Latin, 'at another time, otherwise.'

a•li•as•ing /'ālēəsiNG/ ▸ **n.** in computer graphics, the jagged, or sawtoothed appearance of curved or diagonal lines on a low-resolution monitor.

A-life /'ā ,līf/ ▸ **n.** short for artificial life, the production or action of computer programs or computerized systems that stimulate the characteristics of living organisms: *Not surprisingly, the mimetic potentials of A-life are finding application in the arts, most notably in the emerging field of interactive art.*

al•pha•nu•mer•ic /,ælfən(y)o͞o'merik/ ▸ **adj.** consisting of or using both letters and numerals: *alphanumeric data* | *an alphanumeric keyboard.*

▸ **n.** a character that is either a letter or a number.

DERIVATIVES **al•pha•nu•mer•i•cal adj.**

ORIGIN 1950s: blend of *alphabetical* and *numerical*.

al•pha test /ˈælfə ˌtest/ ▸ **n.** a trial of machinery, software, or other products carried out by a developer before a product is made available for beta testing.

▸**v.** (**a•lpha-test**) [trans.] subject (a product) to a test of this kind: *a new version of an operating system may be alpha tested at our office.*

Alt /ôlt/ ▸ **n.** short for **ALT KEY**.

Alt key /ˈôlt ˌkē/ ▸ **n.** a key on a keyboard that when pressed at the same time as another key gives the second key an alternative function.

ORIGIN late 20th cent.: abbreviation of *alt(ernative) key.*

an•a•log /ˈænlˌôg/ (also **an•a•logue**) ▸ **adj.** relating to or using signals or information represented by a continuously variable physical quantity such as spatial position or voltage. Often contrasted with **DIGITAL**.

ORIGIN early 19th cent.: from French, from Greek *analogon*, neuter of *analogos* 'proportionate.'

an•a•log-to-dig•it•al con•vert•er /ˈænlˌôg tə ˈdijitl kənˈvərtər/ (abbr.: **ADC**) ▸**n.** a device for converting analog signals to digital form.

AND /ænd/ ▸ **n.** a Boolean operator that evaluates as true if and only if all the arguments are true and otherwise has a value of false. See usage note at **BOOLEAN**.

an•i•ma•tion /ˌænəˈmāSHən/ (also **com•put•er an•i•ma•tion** /kəm ˈpyo͞otər anəˌmāSHən/) ▸ **n.** the manipulation of electronic images by means of a computer in order to create moving images.

ORIGIN mid 16th cent.: from Latin *animatio(n-)*, from *animare* 'instill with life'

a•non•y•mous FTP /əˈnänəməs ˈef ˌtē ˈpē/ ▸ **n.** part of the file transfer protocol (FTP) on the Internet that lets anyone log on to an FTP server, using a general username and without a password.

an•ti•a•li•as•ing /ˌæntēˈālēəsiNG; ˌæntī-/ ▸**n.** the reduction of jagged edges on diagonal lines in digital images. [often as modifier] *by combining with the antialiasing function, it enables a smooth and high-speed drawing which was not achievable with previous software processing.*

DERIVATIVES **an•ti•a•li•as v.**

an•ti•vi•rus /ˌæntēˈvīrəs/ ▸ **adj.** [attrib.] (of software) designed to detect and destroy computer viruses.

API ▸ **abbr.** application programming interface.

app /æp/ ▸ **n.** short for **APPLICATION.**

ap•plet /ˈæplit/ ▸ **n.** a very small application, esp. a utility program performing one or a few simple functions.

ORIGIN 1990s: blend of **APPLICATION** and *-let.*

ap•pli•ca•tion /ˌæpliˈkāSHən/ ▸ **n.** a program or piece of software designed and written to fulfill a particular purpose of the user: *a database application.*

ORIGIN Middle English: via Old French from Latin *applicatio(n-),* from the verb *applicare.*

ap•pli•ca•tion pro•gram /ˌæpliˈkāSHən ˌprōgram/ ▸ **n.** another term for **APPLICATION.**

ap•pli•ca•tion pro•gram•ming in•ter•face /ˌæpliˈkāSHən ˌprōgræmiNG ˌintərfās/ ▸ **n.** a system of tools and resources in an operating system, enabling developers to create software applications.

ar•chi•tect /ˈärkiˌtekt/ ▸ **v.** [trans.] (usu. **be architected**) design and make: *few software packages were architected with Ethernet access in mind.*

ORIGIN mid 16th cent.: from French *architecte,* from Italian *architetto,* via Latin from Greek *arkhitektōn,* from *arkhi-* 'chief' + *tektōn* 'builder.'

ar•chi•tec•ture /ˈärkiˌtekCHər/ ▸ **n.** the conceptual structure and logical organization of a computer or computer-based system: *a client/server architecture.*

DERIVATIVES **ar•chi•tec•tur•al** /ˌärkiˈtekCHərəl/ **adj.**; **ar•chi•tec•tur•al•ly adv.**

ORIGIN mid 16th cent.: from Latin *architectura,* from *architectus.*

ar•chive /ˈärkīv/ ▸ **n.** a complete record of the data in part or all of a computer system, stored on an infrequently used medium.

▸ **v.** create an archive of (computer data): [intrans.] *we began archiving in June* | [trans.] *neglecting to archive our files was a costly oversight.*

DERIVATIVES **ar•chi•val** /ärˈkīvəl/ **adj.**

ar•gu•ment /ˈärgyəmənt/ ▸ **n.** a value or address passed to a procedure or function at the time of call.

ar•ith•met•ic log•ic u•nit /'ariTH'metik 'läjik ˌyōōnit/ ▸ **n.** a unit in a computer that carries out arithmetic and logical operations.

ar•ti•fi•cial in•tel•li•gence /ˌärtə'fiSHəl in'telijəns/ (abbr.: **AI**) ▸ **n.** the theory and development of computer systems able to perform tasks that normally require human intelligence, such as visual perception, speech recognition, decision-making, and translation between languages.

ar•ti•fi•cial life /'ärtəˌfiSHəl 'līf/ ▸ **n.** see **A-life.**

ASCII /'æskē/ ▸ **n.** a set of digital codes representing letters, numerals, and other symbols, widely used as a standard format in the transfer of text between computers.

ORIGIN acronym from *American Standard Code for Information Interchange.*

ASP ▸ **abbr.** application service provider, a company providing Internet access to software applications that would otherwise have to be installed on individual computers.

as•sem•ble /ə'sembəl/ ▸ **v.** [trans.] translate (a program) from assembly language into machine code.

ORIGIN Middle English: from Old French *asembler*, based on Latin *ad-* 'to' + *simul* 'together.'

as•sem•bler /ə'semblər/ ▸ **n.** a program for converting instructions written in low-level symbolic code into machine code.

■ another term for **ASSEMBLY LANGUAGE.**

as•sem•bly /ə'semblē/ ▸ **n.** [usu. as adjective] the conversion of instructions in low-level code to machine code by an assembler.

ORIGIN Middle English: from Old French *asemblé*, feminine past participle of *asembler* (see **ASSEMBLE**).

as•sem•bly lan•guage /ə'semblē ˌlæNGgwidj/ ▸ **n.** a low-level symbolic code converted by an assembler.

as•so•ci•a•tive /ə'sōSH(ē)ətiv/ ▸ **adj.** [attrib.] of or denoting computer storage in which items are identified by content rather than by address.

a•syn•chro•nous /ā'siNGkrənəs/ ▸ **adj.** of or requiring a form of computer control timing protocol in which a specific operation begins upon receipt of an indication (signal) that the preceding operation has been completed.

DERIVATIVES **a•syn•chro•nous•ly adv.**

at•tach•ment /ə'tacHmənt/ ▸ **n.** a computer file appended to an e-mail: *the law firm's investigation was expanded a few weeks ago to include whether Stubblefield broke any city policies or copyright laws when he sent an e-mail to city department heads that included an attachment of an MP3 song file.*

au•di•o /'ôdēˌō/ ▸ **n.** sound, esp. when recorded, transmitted, or reproduced: *the machine can retrieve and play audio from a CD-ROM.*

ORIGIN 1930s: independent usage of *audio-*.

au•then•ti•cate /ô'THentiˌkāt/ ▸ **v.** [intrans.] (of a user or process) have one's identity verified.

DERIVATIVES **au•then•ti•ca•tion** /ôˌTHenti'kāSHən/ **n.**; **au•then•ti•ca•tor** /-ˌkāt̲ər/ **n.**

ORIGIN early 17th cent.: from medieval Latin *authenticat-* 'established as valid,' from the verb *authenticare*, from late Latin *authenticus* 'genuine.'

au•thor•ing /'ôTHəriNG/ ▸ **n.** the creation of programs and databases for computer applications such as computer-assisted learning or multimedia products: [as adjective] *an authoring system.*

au•to•com•plete /ˌôtōkəm'plēt/ ▸ **n.** a software function that gives users the option of completing words or forms by a shorthand method on the basis of what has been typed before: *it would allow me to write plug-ins that hook into the editing process so that features like autocomplete, spell-checking, and other niceties could be added.*

▸**v.** [trans.] complete (a word or form) in this way.

DERIVATIVES **au•to•com•ple•tion** **n.** /ˌôtōkəm'plēSHən/

au•to•dial /'ôtōˌdī(ə)l/ ▸ **n.** a function of telephonic equipment that allows for automatic dialing of preprogrammed or of randomly selected numbers: *have a telephone with autodial by your bed.*

▸**v.** (**-dialed**, **-dial•ing**; Brit. **-dialled**, **-dial•ling**) [intrans.] automatically dial a telephone number, with or without human prompting: *the first time I discovered it had autodialed and been on-line for over 2 hours.*

AWK /ôk/ ▸ **n.** a computer programming language used to manipulate large configuration files programmatically.

ORIGIN 1978: from the initials of its creators: Aho, Weinberger, and Kernighan.

AYT ▸ **abbr.** informal (in e-mail or chatrooms) are you there?

B

back•bone /ˈbæk,bōn/ ▸ **n.** a large transmission line or system for carrying data, that local lines or networks connect to.

back end /ˈbæk ˈend/ ▸ **adj.** [attrib.] denoting a subordinate processor or program, not directly accessed by the user, which performs a specialized function on behalf of a main processor or software system: *a back-end database server.*

back•ground /ˈbæk,grownd/ ▸ **n.** used to describe tasks or processes running on a computer that do not need input from the user: *programs can be left running in the background* | [as adjective] *background processing.*

back•lit /ˈbæk,lit/ ▸ **adj.** (of a display screen) illuminated from behind: *a backlit LCD screen.*

back•slash /ˈbæk,slæSH/ ▸ **n.** a backward-sloping diagonal line (\), used in computer commands.

back•space /ˈbæk,spās/ ▸ **n.** a key on computer keyboard that causes the cursor to move backward.

▸ **v.** [intrans.] move a computer cursor back one or more spaces.

back•up /ˈbæk,əp/ ▸ **n.** the procedure for making extra copies of data in case the original is lost or damaged: *automatic online backup* | [as adjective] *a backup disk.*

■ a copy of this type.

back up /ˈbæk ˈəp/ ▸ **v. back something up** make a spare copy of data or a disk.

back•ward com•pat•i•ble /ˈbækwərd kəmˈpætəbəl/ (also **back•wards com•pat•i•ble** /-wərdz/) ▸ **adj.** (of computer hardware or software) able to be used with an older piece of hardware or software without special adaptation or modification.

DERIVATIVES **back•ward com•pat•i•bil•i•ty** /kəm,pætəˈbilitē/ **n.**

band•width /ˈbænd,widTH/ ▸ **n.** the data transmission capacity of a computer network or other telecommunication system.

bang /bæNG/ ▸ **n.** the character "!"

ban•ner /ˈbænər/ (also **ban•ner ad** /ˈbænər ,æd/) ▸ **n.** an advertisement appearing across the top of a web page: *to get a new banner now, click Step 1.* | [as modifier] *advertise and promote your site on thousands of web sites all around the world for free utilizing our award winning banner exchange engine and free web tools!*

BASIC /ˈbāsik/ ▸ **n.** a simple high-level computer programming language that uses familiar English words, designed for beginners and formerly widely used on desktop computers.

ORIGIN 1960s: acronym from *Beginners' All-purpose Symbolic Instruction Code.*

batch /bæCH/ ▸ **n.** a group of records processed as a single unit, usually without input from a user.

batch file /ˈbæCH ,fīl/ ▸ **n.** a computer file containing a list of instructions to be carried out in turn.

batch proc•ess•ing /ˈbæCH ,präsesiNG/ ▸ **n.** the processing of previously collected jobs in a single batch.

baud /bôd/ ▸ **n.** (pl. same or **bauds**) a unit used to express the speed of transmission of electronic signals, corresponding to one information unit or event per second.

■ a unit of data transmission speed for a modem of one bit per second (in fact there is usually more than one bit per event).

ORIGIN 1930s: coined in French from the name of J. M. E. *Baudot* (1845–1903), French engineer who invented a telegraph printing system.

bay /bā/ ▸ **n.** a space in a computer cabinet, into which an electronic device can be installed: *a drive bay.*

ORIGIN late Middle English: from Old French *baie*, from *baer* 'to gape,' from medieval Latin *batare*, of unknown origin.

BBS ▸ **abbr.** bulletin board system.

bcc ▸ **abbr.** blind carbon copy, a copy of an e-mail sent to someone whose name and address isn't visible to other recipients.

bench•mark /ˈbenCH,märk/ ▸ **n.** a problem designed to evaluate the performance of a computer system: *Xstones is a graphics benchmark.*

■ a standard or point of reference against which hardware or software may be compared or assessed.

▸ v. [trans.] evaluate or check (hardware or software) by comparison with a standard.

■ [intrans.] evaluate or check something in this way: *we continue to benchmark.* ■ [intrans.] show particular results during a benchmark test: *the device should benchmark at between 100 and 150 MHz.*

bench•mark test /ˈbenCHˌmärk ˌtest/ ▸ **n.** a test using a benchmark to evaluate a computer system's performance.

bench test /ˈbenCH ˌtest/ ▸ **n.** a test carried out on a machine, a component, or software before it is released for use, to ensure that it works properly.

▸ v. (**bench-test**) [trans.] run a bench test on (something): *they are offering you the chance to bench-test their applications.*

■ [intrans.] give particular results during a bench test: *it bench-tests two times faster than the previous version.*

be•ta test /ˈbātə ˌtest/ ▸ **n.** a trial of machinery, software, or other products, in the final stages of its development, carried out by a party unconnected with its development.

▸ v. (**be•ta-test** /ˈbātə ˈtest/) [trans.] subject (a product) to such a test: *we expect to beta test the model during December.*

bi•di•rec•tion•al /ˌbīdiˈrekSHənl/ ▸ **adj.** functioning in two directions: *bidirectional audio and video connections.*

big en•di•an /ˈbig ˈendēən/ ▸ see ENDIAN.

bi•na•ry code /ˈbīnərē ˌkōd/ ▸ **n.** a coding system using combinations of the binary digits 0 and 1 to represent a letter, digit, or other character in a computer or other electronic device.

bi•o•com•pu•ter /ˈbīōkəmˌpyo͞otər/ ▸ **n.** a computer based on circuits and components formed from biological molecules or structures that would be smaller and faster than an equivalent computer built from semiconductor components.

■ a human being, or the human mind, regarded as a computer.

bi•o•com•put•ing /ˌbīōkəmˈpyo͞otiNG/ ▸ **n.** the design and construction of computers using biochemical components: *while biocomputing includes ways to do rudimentary computing with DNA itself, scientists have begun looking at ways to do computations in whole* *cells by engineering part of the cells' DNA and the machinery controlled by those genes.*

■ an approach to programming that seeks to emulate or model biological processes. ■ computing in a biological context or environment.

BIOS /ˈbīōs/ ▸ **n.** a set of computer instructions in firmware that control input and output operations.

ORIGIN acronym from *Basic Input-Output System.*

bit /bit/ ▸ **n.** a unit of information expressed as either a 0 or 1 in binary notation.

ORIGIN 1940s: blend of *binary* and *digit.*

bit•map /ˈbitˌmæp/ ▸ **n.** a representation in which each item corresponds to one or more bits of information, esp. the information used to control the display of a computer screen.

▸ **v.** (**bit•mapped, bit•map•ping**) [trans.] represent (an item) as a bitmap.

BITNET /ˈbitˌnet/ (also **Bit•net**) **trademark** a data transmission network founded in 1981 to link North American academic institutions and to interconnect with other information networks.

bleed•ing edge /ˈblēdiNG ˈej/ ▸ **n.** the very forefront of technological development: *a design that many people believe is still too bleeding edge for large mission-critical systems.*

ORIGIN 1980s: on the pattern of *leading edge, cutting edge.*

bloat•ware /ˈblōtˌwe(ə)r/ ▸ **n.** informal software that requires an amount of disk storage space that is grossly incommensurate with its utility: *none of the programs on this page is bloatware, so they can be downloaded fairly quickly.*

BLOB /bläb/ ▸ **n.** binary large object; a stored block of data without reference to the database management system except size and location.

ORIGIN acronym.

blog /bläg/ informal ▸ **n.** a weblog.

▸ **v.** (**blogged, blog•ging**) [intrans.] add new material to or regularly update a weblog.

DERIVATIVES **blog•ger n.**

blog•o•sphere /ˈbläɡəˌsfi(ə)r/ ▸ **n.** the world of weblogs: *the blog's* *dullness was inspired—if that is the correct word—by Mr. Walker's careful study of the blogosphere.*

Blue•tooth /ˈblo͞oˌto͞oTH/ ▸ **trademark** a standard for the short-range wireless interconnection of computers, cell phones, and other electronic devices.

ORIGIN 1990s: said to be named after King Harald *Bluetooth* (910–985), credited with uniting Denmark and Norway, as Bluetooth technology unifies the telecommunications and computing industries.

board /bôrd/ ▸ **n.** a flat insulating sheet used as a mounting for an electronic circuit: [with adjective] *a graphics board.*

■ another term for **EXPANSION CARD.**

ORIGIN Old English *bord*, of Germanic origin; related to Dutch *boord* and German *Bort*; reinforced in Middle English by Old French *bort* 'edge, ship's side' and Old Norse *borth* 'board, table.'

book•mark /ˈbo͝okˌmärk/ ▸ **v.** [trans.] make a record of (the address of a file, Internet page, etc.) to enable quick access by a user: *its database pool is expected to grow over time, and is well worth bookmarking.*

▸ **n.** any of a collection of Web site addresses saved in a computer file for easy retrieval: *you can access any work instantly from the main menu and designate up to 99 bookmarks.*

Bool•e•an /ˈbo͞olēən/ ▸ **adj.** denoting a system of algebraic notation used to represent logical propositions, esp. in computing and electronics: *a Boolean search.*

▸ **n.** a binary variable, having two possible values called "true" and "false."

ORIGIN mid 19th cent.: from the name of English mathematician G. *Boole* + *-an.*

USAGE: Using Boolean search operators AND, OR, and NOT can significantly improve the quality of your Web searches. For instance, searching with the two words *unicycle OR juggling* will get you pages where either term is mentioned alone. The search stops looking if it finds either term, so the *unicycle* pages might contain the word *juggling*, and might not. To find only pages where both words are present, use *unicycle AND juggling*. To find only pages about unicycles that don't mention juggling, use *unicycle NOT juggling.*

boot /bo͞ot/ ▸ **n.** (also **boot up** /ˈbo͞ot ˌəp/) [usu. as adjective] the process of starting a computer and putting it into a state of readiness for operation: *the boot sector of the hard disk is referred to every time the PC is turned on.*

▸**v.** [trans.] start (a computer) and put it into a state of readiness for operation: *the menu will be ready as soon as you* ***boot up*** *your computer* | [intrans.] *the system won't* ***boot from*** *the hard drive.*

ORIGIN from **BOOTSTRAP**.

boot•a•ble /ˈbo͞otəbəl/ ▸ **adj.** (of a disk) containing the software required to boot a computer.

boot•strap /ˈbo͞otˌstræp/ ▸ **n.** a technique of loading a program into a computer by means of a few initial instructions that enable the introduction of the rest of the program from an input device.

boot-up /ˈbo͞ot ˌəp/ ▸ **n.** see **BOOT**.

bot /bät/ ▸ **n.** an autonomous program on a network (esp. the Internet) that can interact with computer systems or users, esp. one designed to respond or behave like a player in an adventure game. ■ a computer program that behaves like a human user in some specific capacity: *often, a bot looks like any other human user, and you might not be able to tell you're hanging out with software.*

ORIGIN 1980s: shortening of *robot*.

-bot ▸ **combining form** used to form nouns denoting a computer program or robot with a very specific function: *she wanted to direct the female's behavior. She envisioned controlling a robot that could replace the female and fool a male completely. So, she created the fembot.*

box /bäks/ ▸ **n.** informal **1** a personal computer or workstation: *I'm having networking problems where windows boxes can route through my Linux gateway, but my Linux boxes can't.* ■ a casing containing a computer.

2 an area on a computer screen for user input or displaying information: *a dialog box.*

PHRASES **(right) out of the box** describing a newly purchased product that works immediately, without any special assembly or training: *a completely preconfigured system you can quickly install right out of the box.*

ORIGIN late Old English, probably from late Latin *buxis*, from Latin *pyxis* 'boxwood box,' from Greek *puxos*).

bpi ▸ **abbr.** bits per inch, used to indicate the density of data that can be stored on magnetic tape or similar media.

bps ▸ **abbr.** bits per second.

break•point /ˈbrākˌpoint/ ▸ **n.** (also **break point**) a place in a computer program where the sequence of instructions is interrupted, esp. by another program or by the operator.

broad•band /ˈbrôdˌbænd/ ▸ **adj.** of or using signals over a wide range of frequencies in high-capacity telecommunications, esp. as used for access to the Internet: *broadband access.*

▸**n.** signals over such a range of frequencies: *wireless broadband* | *the ability to uplink on broadband.*

bro•chure•ware /brōˈsʜo͝orˌwe(ə)r/ ▸ **n.** Web sites or Web pages produced by converting a company's printed marketing or advertising material into an Internet format, typically providing little or no opportunity for interactive contact with prospective customers: *certainly the wealth of brochureware on the Internet makes comparison shopping there much easier than physically visiting or even telephoning each vendor.*

brown goods /ˈbrown ˌgo͝odz/ ▸ **plural n.** computers, television sets, audio equipment, and similar household appliances: *our supply chain needs to accommodate highly perishable products, as well as white and brown goods.*

browse /browz/ ▸ **v.** [trans.] read or survey (data files) stored on a disk drive or a network.

brows•er /ˈbrowzər/ ▸ **n.** a program with a graphical user interface for displaying HTML files, used to navigate the World Wide Web: *a Web browser.*

bub•ble•jet print•er /ˈbəbəlˌjet ˌprintər/ ▸ **n.** a kind of inkjet printer.

bud•dy list /ˈbədē ˌlist/ ▸ **n.** a list of people who you communicate with over the Internet using instant messaging.

buff•er /ˈbəfər/ ▸ **n.** a temporary memory area or queue used when transferring data between devices or programs operating at different speeds.

bug /bəg/ ▸ **n.** an error in a computer program or system.

bug•gy /ˈbəgē/ ▸ **adj.** (**bug•gi•er**, **bug•gi•est**) (of a computer program or system) faulty in operation.

build /bild/ ▸ **v.** (past and past part. **built** /bilt/) [trans.] (often **be built**) compile (a program, database, index, etc.).

■ [intrans.] (of a program, database, index, etc.) be compiled.

▸ **n.** a compiled version of a program.

■ the process of compiling a program.

bul•le•tin board /ˈbo͝olitn ˌbôrd/ (also **bul•le•tin board sys•tem** /ˈbo͝olitn ˌbôrd ˌsistəm/) ▸ **n.** an information storage system designed to permit any authorized computer user to access and add to it. Compare with **MESSAGE BOARD**.

bun•dle /ˈbəndl/ ▸ **n.** a set of software, hardware, or services sold together.

▸ **v.** [trans.] sell (hardware, software, support services) as a package: *imaging software comes bundled with the scanner.*

ORIGIN Middle English: perhaps originally from Old English *byndelle* 'a binding,' reinforced by Low German and Dutch *bundel* (to which *byndelle* is related).

burn /bərn/ ▸ **v.** (past and past participle **burned** or **burnt**) [trans.] **1** produce (a compact disc) by copying from an original or master copy. **2** copy data to (a CD or DVD) from a CD, DVD, audio or video file stored on a drive, or other digital source.

ORIGIN A high-intensity laser is used to heat, or 'burn,' a layer of dye within the disk to create the copy.

burn•er /ˈbərnər/ ▸ **n.** a drive for encoding a compact disc or DVD by creating nonreflective areas on the disk with a high-intensity laser.

burn•out /ˈbərnˌowt/ ▸ **n.** failure of an electrical device or component through overheating: [with adjective] *that brand of monitor is prone to pixel burnout.*

bus /bəs/ ▸ **n.** (pl. **bus•es** or **bus•ses**) a distinct set of conductors carrying data and control signals within a computer system, to which pieces of equipment may be connected.

byte /bīt/ ▸ **n.** a group of binary digits or bits (usually eight) operated on as a unit. Compare with **BIT**.

■ such a group as a unit of memory size.

ORIGIN 1960s: an arbitrary formation based on *bit* and *bite*.

WHICH OPERATING SYSTEM IS RIGHT FOR YOU?

It's the age-old computer question, approached with the same intensity as politics and sports: which operating system (OS for short) is right for you?

But ask yourself a more fundamental question first: what will you use your computer for? Is it casual word processing and e-mail? The Great American Novel? Number crunching? The Internet? Recreational gaming? Video editing? Graphic design?

The point is this: how you use your computer determines what software you need, and the availability of that software for a given operating system in turn guides your choice of OS. But remember, an operating system is linked to a specific kind of computer. For the home buyer, that means, essentially, a choice between a Mac and a PC. And your comfort level, as you try out different user interfaces, will ultimately have a profound effect on your choice of computer and operating system.

Broadly, there are three popular OS families currently available: Windows, Macintosh OS's, and countless flavors of Linux.

Windows—If we include versions from Win 95 on, Microsoft Windows has the largest present and potential market share—so much so that many third-party software companies simply cannot afford to develop programs for any other OS. This alone is reason enough for many people to choose it. But others point out that just because 100 million cows eat grass, that doesn't mean that they should eat it too. For those dissident cows among us, there are indeed some conspicuous cons to consider: Windows (particularly in its older incarnations—95 through ME) is notorious for its frequent crashes as well as its dangerous security holes, with Microsoft having to release dozens of security patches annually. And nearly all of those nasty computer viruses, trojans, and worms that you hear about are written to exploit its weaknesses. In addition, a Windows PC generally requires more technical support over its lifetime than an Apple Macintosh.

On the other hand, there are pros to balance the cons. Recent versions of the OS (2000, XP, and beyond) are much more reliable, Microsoft promises updates and long-term support, and—more important—Windows systems are compatible with the largest variety of the latest software applications (from office suites, reference works, and utilities to vast quantities of games) as well as with the newest peripherals. Technical support personnel are more likely to be familiar with Windows PCs, in which it is relatively easy to install upgraded hardware components. And if you want speed, desktop PCs give you more gigahertz for the buck—provided you do your research before buying.

Macintosh—Apple has been encouraging Mac owners to upgrade to Macintosh OS X (pronounced oh-ess-ten; that's a Roman numeral) since its introduction in 2001, having dropped development of Macintosh OS 9 (that's an *Arabic* numeral; hence the confusion). The practical advantages of OS X derive from its Unix underlayer. Unix software, freely developed and tweaked by multiple programmers for more than 30 years, is server-oriented, targeted at true computer geeks, and designed around open, or public, standards, not closed, or proprietary ones. This not only makes for better compatibility with other platforms, it also makes OS X more stable than its predecessors. In addition, the OS comes with a wealth of genuine, up-to-date Unix tools, thereby becoming not only an easy entrée for novices (for which the Mac has always been famous) but also a productive platform for computer experts. Additionally, it runs older OS 9 programs and has achieved, in its yearly revisions, an increasingly attractive interface. Apple has always been known for beauty, and its computers—the machines themselves—are usually more stylish, more avant garde, than their Windows counterparts. The Mac as exemplar of the leading edge doesn't stop there; it has often been the first popular platform to offer new technologies, from the early Graphical User Interface to the more recent Firewire and Wi-Fi.

But there are drawbacks to entering the Apple world. Although, feature for feature, you pay about the same for an Apple laptop as for a PC laptop, desktop Macintosh units are comparatively pricey. And because Apple's market share is smaller than that of Windows, software for its OS is released later, if at all. Good games, for example, are rare and late. But in three areas, the Mac has traditionally been king: design, design, design. High-end graphics design software emerged first for the Mac, as did video- and photo-editing programs and desktop publishing applications. Both the machine and the operating system are still favored by those whose work demands creative artistry.

Linux and Unix—The odds are pretty good that you're not going to buy a Sun server in order to run Unix at home, but if you're curious (or strapped for cash) and technically inclined, you can experiment by finding a cheap PC on eBay or at a garage sale and install a free operating system. A good choice for that is the Unixlike Linux. There are many distributions of Linux, known by the cognoscenti as "distros." They come bundled with thousands of programs, with thousands more available for download. It's all pretty overwhelming, and Linux is not for the fainthearted. Installing and using it requires patience, technical skill, and lots of time. You may spend more time tweaking and troubleshooting the system than you do using it for day-to-day tasks. Yet once a Linux system is running, its power is incredible and undeniable. There's a world of tempting open-source and free software that can cover most of your day-to-day needs—word processors, spreadsheets, MP3 players, web browsers, and programming tools. And almost all of it is completely free, yours for the trouble of downloading, compiling, and installing it. Some of the Linux distros (such as Knoppix) offer CD-bootable versions that will let you dip your toe into the Linux world without committing to it. In other words, you can run the OS and its compatible applications directly from the CD without ever installing it on your computer! Also worth considering, if you're of a technical bent, are OpenBSD and NetBSD, which are true Unix operating systems.

There are a few other points to consider when choosing an operating system:

1) If you're replacing an old computer, you will probably want a new one with the same operating system; you really don't want to replace all your software too! Windows programs don't run on a Mac without an add-on program such as VirtualPC (which runs slowly) and Mac programs don't run on a Windows computer at all. But it's not all bad; file formats of the major programs that run on both Macs and PCs are almost universally cross-platform, so you may well be able to preserve your data. A Word document created on a Mac can be opened on a PC and vice-versa, as long as the versions of Word on the two computers are compatible. Keep in mind that while files created with older versions of a program are generally readable by newer ones, there's no guarantee that files created with new versions will be readable by older ones.

2) If you take work home from the office, it may be important to have as little difference as possible between your home and office computers. In addition, some companies have site-licensing agreements that may legally permit your company's IT department to install work-related office software on your home computer. That can save you a bundle.

3) On the other hand, you may want to learn to use more than one operating system, explore a variety of software pro-

grams, and plumb the mysteries of the computers that run them. Expanding your skills by getting a home computer different from the one in your office can conceivably make you more attractive in the job market.

4) Finally, if you have a willing friend or relative who is skilled at troubleshooting computers, you might want to use the operating system that person uses. You'll get your own personal tech support department! (But don't overdo it; buy an appropriate book and read manuals and help files.)

In sum, for casual computing, either a Mac with OS X or a PC with a new Windows OS is just fine. For bringing work home or doing your doctoral dissertation, think compatibility with your office or school. For gaming, superfast PCs beckon (you may eventually want to build your own!), and you'll most likely want the newest version of Windows. For budding techno-nerds and the Windows-weary, Linux is waiting. For working in desktop publishing or high-end graphics, you'll probably want to emulate most of your peers and get a Mac, although much of that software is now available for Windows as well. And for running your own business, let software be your guide.

C

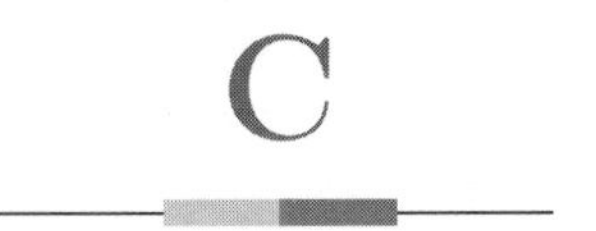

C /sē/ (also **c**) ▸ **n.** (pl. **Cs** or **C's**) a high-level computer programming language originally developed for implementing the UNIX operating system. [ORIGIN: formerly known as *B*, abbreviation of *BCPL.*]

ca•ble mo•dem /ˈkābəl ˌmōdəm/ ▸ **n.** a type of modem that connects a computer or local network to broadband Internet service through the same cable that supplies cable television service: [as adjective] *a cable-modem connection.*

■ the service connection made via a cable modem: *a broadband Internet connection, such as DSL or cable modem.*

cache /kæsh/ (also **cache mem•o•ry** /ˈkæsh ˌmem(ə)rē/) ▸ **n.** an auxiliary memory from which high-speed retrieval is possible.

▸ **v.** [trans.] store (data) in a cache memory: *the ability to cache images in memory.*

■ provide (hardware) with a cache memory.

ORIGIN late 18th cent.: from French, from *cacher* 'to hide.'

CAD /kæd/ ▸ **abbr.** computer-aided design.

CADCAM (also **CAD/CAM**) /ˈkædˈkæm/ ▸ **abbr.** computer-aided design, computer-aided manufacturing.

CAE ▸ **abbr.** computer-aided engineering.

CAI ▸ **abbr.** computer-assisted (or -aided) instruction.

CAL /ˈsē ˈā ˈel; kæl/ ▸ **abbr.** computer-assisted (or -aided) learning.

call /kôl/ ▸ **v.** [trans.] cause the execution of (a subroutine).

ORIGIN late Old English *ceallian*, from Old Norse *kalla* 'summon loudly.'

call•back /ˈkôlˌbæk/ ▸ **n.** a security feature used by systems accessed by telephone, in which a remote user must log on using a previously registered phone number, to which the system then places a return call.

CAM /kæm/ ▸ **abbr.** computer-aided manufacturing.

cam•boy /'kam,boi/ ▸ **n.** a boy or man who poses for a webcam.

cam•girl /'kam,gərl/ ▸ **n.** a girl or woman who poses for a webcam: *many camgirls and camboys say they haven't revealed to their parents the full extent of their online conversations.*

Can•a•darm /'kanə,därm/ ▸ **n.** the popular name for a robotic manipulation system designed for use in zero gravity; it has accompanied numerous space missions as a component on space shuttles.

ORIGIN 1970s: blend of *Canada* (where it was manufactured) and *arm.*

can•cel•bot /'kænsəl,bät/ ▸ **n.** a program that searches for and deletes specified mailings from Internet newsgroups. Also see **BOT**.

ORIGIN 1990s: from *cancel* + **BOT**.

ca•pa•bil•i•ty /,kāpə'bilitē/ ▸ **n.** (pl. **ca•pa•bil•i•ties**) [usu. with adjective] a facility on a computer for performing a specified task: *a graphics capability.*

cap•ture /'kæpCHər/ ▸ **v.** [trans.] cause (data) to be stored in a computer: *capture an image to the server.*

ORIGIN mid 16th cent.: from French, from Latin *captura*, from *capt-* 'seized, taken,' from the verb *capere.*

card /kärd/ ▸ **n.** short for **EXPANSION CARD**.

card read•er /'kärd ,rēdər/ ▸ **n.** an electronic device that reads and transfers data from various portable memory storage devices: *if your computer doesn't have a memory card slot, companies such as Lexar offer accessory card readers that plug into your computer and let you download data from memory cards.*

car•riage re•turn /'kærij ri,tərn/ ▸ **n.** another term for **RETURN**.

case mod /'kās ,mäd/ ▸ **n.** another term for **MOD**.

case-sen•si•tive /'kās 'sensitiv/ ▸ **adj.** (of a program or function) differentiating between capital and lowercase letters: *the password system is case-sensitive.*

■ (of input) treated differently depending on whether it is in capitals or lowercase text.

CAT /'kæt/ ▸ **abbr.** ■ computer-assisted (or -aided) testing.

CD ▸ **abbr.** compact disc.

CD-R ▸ abbr. compact disc recordable, a blank CD on which data, including music, can be permanently recorded.

CD-ROM /ˈsē ˌdē ˈräm/ ▸ n. a compact disc used as a read-only optical memory device for a computer system.

ORIGIN 1980s: acronym from *compact disc read-only memory*.

CD-RW ▸ abbr. compact disc rewritable, a blank CD that can be recorded, erased, and rerecorded many times

■ a disk drive that can read and record CDs

cel•lu•lar au•tom•a•ton /ˈselyələr ôˈtämətən/ ▸ n. (pl. **cel•lu•lar au•tom•a•ta** /ˈselyələr ôˈtämətə/) one of a set of units in a mathematical model that have simple rules governing their replication and destruction. They are used to model complex systems composed of simple units such as living things or parallel processors.

cen•tral proc•ess•ing u•nit /ˈsentrəl ˈpräsesiNG ˌyo͞onit/ (also **cen•tral proc•es•sor**) ▸ n. see **CPU**.

cgi ▸ abbr. common gateway interface, a script standard for writing interactive programs generated by visitors to web pages, such as forms and searches.

cgi-bin /ˈsēˌjēˈī ˌbin/ ▸ n. a server directory where cgi programs are stored: [as modifier] *cgi-bin files*.

chad /CHæd/ ▸ n. a piece of waste material created by punching cards or tape used for computer data input.

ORIGIN 1950s: of unknown origin.

chang•er /ˈCHānjər/ ▸ n. a device that holds several computer disks or compact disks and is able to switch between them.

char•ac•ter /ˈkærəktər/ ▸ n. a printed or written letter or symbol.

■ a symbol representing a letter or number. ■ the bit pattern used to store such a symbol.

ORIGIN Middle English: from Old French *caractere*, via Latin from Greek *kharaktēr* 'a stamping tool.'

char•ac•ter code /ˈkærəktər ˌkōd/ ▸ n. the binary code used to represent a letter or number: *for convenience, we call such embedded information—and all other aspects of a bit stream's representation, including byte length, character code and structure—the encoding of a document file.*

char•ac•ter rec•og•ni•tion /ˈkærəktər ˌrekəgˌniSHən/ ▸ n. the identification by electronic means of printed or written characters.

char•ac•ter string /ˈkærəktər ˌstriNG/ ▸ **n.** a linear sequence of characters stored in or processed by a computer.

chat room /ˈCHæt ˌro͞om/ ▸ **n.** an area on the Internet or other computer network where users can communicate, typically about a particular topic.

chat•ter•bot /ˈCHatərˌbät/ ▸ **n.** a computer program designed to interact with people by simulating human conversation: *one useful thing Andrette knows how to do is to tell you what movies are playing near you, and even in this simple task the chatterbot became confused because I gave my zip code too early in the process.*
ORIGIN 1990s: blend of *chatter* and *(ro)bot.*

check•box /ˈCHekˌbäks/ ▸ **n.** a small box on a computer screen that, when selected by the user, is filled with a check mark or an X to show that the feature described alongside it has been enabled.

check•sum /ˈCHekˌsəm/ ▸ **n.** an error checking device consisting of a number automatically generated from data in such a way that any change to the data results in a change to the number: *the first record header contains the checksum of the login ID of the local administrator.*

chip /CHip/ ▸ **n.** an integrated circuit.
■ a wafer of semiconducting material, usually silicon, used to make an integrated circuit.

chip•set /ˈCHipˌset/ ▸ **n.** a group of integrated circuits that form the set needed to make a computer motherboard.

chy•ron /ˈkīrän/ ▸ **n.** trademark an electronically generated caption superimposed on a television or movie screen.
ORIGIN 1970s: The Chyron Corporation, its manufacturer.

clean room /ˈklēn ˌro͞om; ˌro͝om/ ▸ **n.** an environment free from dust and other contaminants, used chiefly for the manufacture of electronic components: *Daw Technologies Inc. said it received a contract valued at more than $5.7 million to build a clean room for a semiconductor manufacturer.*

click /klik/ ▸ **n.** an act of pressing a mouse button.
▸ **v.** [trans.] press (a mouse button): *click the left mouse button twice.*
■ [intrans.] (**click on**) select (an item represented on the screen or a particular function) by pressing one of the buttons on the mouse

when the cursor is over the appropriate symbol: *just click on this link.*

click•a•ble /ˈklikəbəl/ ▸ **adj.** (of text or images on a computer screen) such that clicking on them with a mouse will produce a reaction.

click rate /ˈklik ˌrāt/ (also **click-through rate** /ˈklik ˈTHro͞o ˌrāt/) ▸ **n.** the percentage of people viewing a Web page who access a hypertext link to a particular advertisement: *despite recent layoffs and a decline in the click rate of banner ads, Advertising.com still expects to prosper this year.*

clicks and mor•tar /ˈkliks ən ˈmôrtər/ ▸ **n.** used to refer to a traditional business that has expanded its activities to operate also on the Internet: *clicks and mortar has often been dismissed by analysts who suggest that real world outfits that move online will end up cannibalizing their existing business.*

■ [usually attributive] a business or business model that involves both Internet selling and physical stores: *many investors are optimistic about the company's use of the "clicks and mortar" strategy, in which walk-in customers use a Borders intranet site to search for titles and order books once they are in the store.*

click•stream /ˈklikˌstrēm/ ▸ **n.** a series of mouse clicks made by a user while accessing the Internet, especially as monitored to assess a person's interests for marketing purposes: *clickstreams, as they are called, enjoy a digital afterlife in commercial databases, where raw statistics about our on-line behavior are transformed into useful information and then warehoused for future application, sale or barter.*

click-through /ˈklik ˌTHro͞o/ ▸ **n.** the ratio of clicks that an Internet ad receives to page views of the ad: *a Web site's dynamics should be reflected in the average rate of click-through across many advertising banners.*

cli•ent /ˈklīənt/ ▸ **n.** (in a network) a desktop computer or workstation that is capable of obtaining information and applications from a server.

■ (also **cli•ent ap•pli•ca•tion** or **pro•gram**) a program that is capable of obtaining a service provided by another program.

cli•ent-serv•er /ˈklīənt ˈsərvər/ ▸ **adj.** denoting a computer system in which a central server provides data to a number of networked workstations.

clip /klip/ ▸ v. (**clipped**, **clip•ping**) [trans.] process (an image) so as to remove the parts outside a certain area.

ORIGIN Middle English: from Old Norse *klippa*, probably imitative.

clip art /ˈklip ˌärt/ ▸ n. predrawn pictures and symbols that computer users can add to their documents, often provided with word-processing software and drawing packages.

clip•board /ˈklipˌbôrd/ ▸ n. a temporary storage area where text or other data cut or copied from a file is kept until it is pasted into another file.

clip•per /ˈklipər/ (also **clip•per chip** /ˈklipər ˌCHip/) ▸ n. a microchip that inserts an identifying code into encrypted transmissions, allowing them to be deciphered by a third party having access to a government-held key: *the Clinton Administration has been pushing for the development of the Clipper Chip, a technology that would enable the government to "wiretap" scrambled digital communications.*

clock speed /ˈkläk ˌspēd/ ▸ n. the operating speed of a computer or its processor, defined as the rate at which it performs internal operations and expressed in cycles per second (megahertz).

clone /klōn/ ▸ n. a microcomputer designed to simulate exactly the operation of another, typically more expensive, model: *an IBM PC clone.*

ORIGIN early 20th cent.: from Greek *klōn* 'twig.'

close /klōz/ ▸ v. [trans.] make (a file, program, or function) inactive and properly stored after use until needed again.

ORIGIN Middle English: from Old French *clos-*, stem of *clore*, from Latin *claudere* 'to shut.'

CMYK ▸ abbr. cyan, magenta, yellow, and black, the four colors used in most color printers, usually in two ink cartridges, one of black ink and the other containing cyan, magenta, and yellow inks in separate reservoirs.

ORIGIN the final 'k' in *black* is used to differentiate black from blue; the color scheme RGB (red, green, blue) is used for color computer display screens.

COBOL /ˈkōˌbôl/ ▸ **n.** a computer programming language designed for use in commerce.

ORIGIN 1960s: from *co(mmon) b(usiness) o(riented) l(anguage).*

co•bot /ˈkōˌbät/ ▸ **n.** a computer-controlled robotic device designed to assist a person: *when a cobot and a UAW member work together to install a large truck instrument panel, the human being does the lifting (see how smart the machines really are) while the cobot provides guidance and direction.*

ORIGIN blend of *collaborative* and *robot.*

code /kōd/ ▸ **n. 1** program instructions: *hundreds of lines of code | assembly code.*

2 a system of words, letters, figures, or other symbols used to represent others, esp. for the purposes of secrecy: *a numeric code.*

▸ **v.** [trans.] convert (instructions or information) into code: *properly code non-ASCII text.*

ORIGIN Middle English: via Old French from Latin *codex, codic-.*

code mon•key /ˈkōd ˌməNGkē/ ▸ **n.** slang a computer programmer, especially an inexperienced or plodding one: *a code monkey in the trenches who needs a job to pay the bills isn't necessarily an enemy of open source.*

cod•er /ˈkōdər/ ▸ **n.** a computer programmer: *the flaws could be used by malicious coders to create new worms or "Trojan horse" attacks, but Microsoft said it doesn't believe any hackers have taken advantage of the security flaws.*

cog•ni•tive pros•the•sis /ˈkägnitiv präsˈTHēsis/ ▸ **n.** an electronic computational device that extends the capability of human cognition or sense perception.

col•li•sion /kəˈliZHən/ ▸ **n.** an event of two or more records being assigned the same location in memory.

■ an instance of simultaneous transmission by more than one node of a network.

ORIGIN late Middle English: from late Latin *collisio(n-)*, from Latin *collidere* 'strike together.'

com•bo box /ˈkämbō ˌbäks/ ▸ **n.** informal a type of dialogue box containing a combination of controls, such as sliders, text boxes, and drop-down lists.

com•bo drive /'kämbō ˌdrīv/ ▸ **n.** an optical disk drive that can read and record CDs and can also read DVDs.

com•mand /kə'mænd/ ▸ **n.** an instruction or signal that causes a computer to perform one of its basic functions.

ORIGIN Middle English: from Old French *comander* 'to command,' from late Latin *commandare*, from *com-* (expressing intensive force) + *mandare* 'commit, command.'

com•mand-driv•en /kə'mænd 'drivən/ ▸ **adj.** (of a program or computer) operated by means of commands keyed in by the user or issued by another program or computer.

com•mand lan•guage /kə'mænd 'læNGgwij/ ▸ **n.** a computer programming language composed chiefly of a set of commands or operators, used esp. for communicating with the operating system of a computer.

com•ment /'käment/ ▸ **n.** a piece of text placed within a program to help other users to understand it, which the computer ignores when running the program.

▸ **v.** [trans.] place a piece of explanatory text within (a program) to assist other users.

■ turn (part of a program) into a comment so that the computer ignores it when running the program: *you could try* ***commenting out*** *that line.*

ORIGIN late Middle English: from Latin *commentum* 'contrivance' (in late Latin also 'interpretation'), neuter past participle of *comminisci* 'devise.'

com•pact disc /'kämpækt 'disk/ (also **com•pact disk**) (abbr.: **CD**) ▸ **n.** a small plastic disc on which music or other digital information is stored, and from which the information can be read using reflected laser light. See also **CD-R**, **CD-ROM**, and **CD-RW**.

com•pat•i•ble /kəm'pæṯəbəl/ ▸ **adj.** (of a piece of software, a computer, or other device) able to be used with a specified piece of equipment or software without special adaptation or modification: *the printer is fully compatible with all leading software.*

▸ **n.** a computer that can use software designed for another make or type.

DERIVATIVES **com•pat•i•bil•i•ty** /kəmˌpæṯə'biliṯē/ **n.**

ORIGIN late Middle English: from French, from medieval Latin *compatibilis*, from *compati* 'suffer with.'

com•pile /kəm'pīl/ ▸ **v.** [trans.] (of a computer) convert (a program) into a machine-code or lower-level form in which the program can be executed.

DERIVATIVES **com•pil•er n.**

ORIGIN Middle English: from Old French *compiler* or its apparent source, Latin *compilare* 'plunder or plagiarize.'

comp•ing /'kämpiNG/ ▸ **n.** the process of making composite images electronically.

com•pos•ite /kəm'päzit/ ▸ **v.** [trans.] [usu. as noun] (**compositing**) combine (two or more images) to make a single picture, esp. electronically: *photographic compositing by computer.*

ORIGIN late Middle English (describing a number having more than one digit): via French from Latin *compositus*, past participle of *componere* 'put together.'

com•press /kəm'pres/ ▸ **v.** [trans.] (often **be compressed**) alter the form of (data) to reduce the size of the file.

DERIVATIVES **com•press•i•ble** /-əbəl/ **adj.**

ORIGIN late Middle English: from Old French *compresser* or late Latin *compressare*, frequentative of Latin *comprimere*, from *com-* 'together' + *premere* 'to press'; or directly from *compress-* 'pressed together,' from the verb *comprimere*.

com•pres•sion /kəm'presHən/ ▸ **n.** the process of compressing data.

ORIGIN late Middle English: via Old French from Latin *compressio(n-)*, from *comprimere* 'press together' (see **COMPRESS**).

com•pu•ta•tion /ˌkämpyŏŏ'tāsHən/ ▸ **n.** the use of computers, esp. as a subject of research or study.

■ the action of mathematical calculation: *statistical computations.*

ORIGIN late Middle English: from Latin *computatio(n-)*, from the verb *computare* (see **COMPUTE**).

com•pu•ta•tion•al /ˌkämpyŏŏ'tāsHənl/ ▸ **adj.** using computers: *the computational analysis of English.*

■ of or relating to computers: *computational power.* ■ of or relating to the process of mathematical calculation.

DERIVATIVES **com•pu•ta•tion•al•ly adv.**

com•pute /kəm'pyo͞ot/ ▸ **v.** [trans.] (often **be computed**) determine (something) using a computer: *we can compute the exact increase.*
■ [intrans.] make a calculation using a computer: *modern circuitry can compute faster than any chess player.*
DERIVATIVES **com•put•a•ble adj.**
ORIGIN early 17th cent.: from French *computer* or Latin *computare*, from *com-* 'together' + *putare* 'to settle (an account).'

com•put•er /kəm'pyo͞otər/ ▸ **n.** an electronic device for storing and processing data, typically in binary form, according to instructions in a variable program.

com•put•er an•i•ma•tion /kəm'pyo͞otər ˌænə'māsʜən/ ▸ **n.** see **ANIMATION.**

com•put•er•ese /kəmˌpyo͞otə'rēz; -'rēs/ ▸ **n.** the symbols and rules of a computer programming language.
■ the jargon associated with computers.

com•put•er graph•ics /kəm'pyo͞otər 'græfiks/ ▸ **plural n.** another term for **GRAPHICS.**

com•put•er•ist /kəm'pyo͞otərist/ ▸ **n.** a (frequent) user of computers: *the reasonably prudent computerist has an anti-virus package.*

com•put•er•ize /kəm'pyo͞otəˌrīz/ ▸ **v.** [trans.] [often as adjective] (**com•pu•ter•ized**) convert to a system that is operated or controlled by computer: *the advantages of computerized accounting.*
■ convert (information) to a form that is stored or processed by computer: *a computerized register of dogs.*
DERIVATIVES **com•put•er•i•za•tion** /kəmˌpyo͞otərə'zāsʜən/ **n.**

com•put•er-lit•er•ate /kəm'pyo͞otər ˌlitərit/ ▸ **adj.** (of a person) having sufficient knowledge and skill to be able to use computers; familiar with the operation of computers: *students are more computer-literate than ever.*
DERIVATIVES **com•put•er lit•er•a•cy** /kəm'pyo͞otər ˌlitərəsē/ **n.**

com•put•er pro•gram•mer /kəm'pyo͞otər ˌprōgræmər/ ▸ **n.** a person who writes programs for the operation of computers, esp. as an occupation.

com•put•er sci•ence /kəm'pyo͞otər ˌsīəns/ ▸ **n.** the study of the principles and use of computers.

com•put•er vi•rus /kəm'pyo͞otər ˌvīrəs/ ▸ **n.** see **VIRUS.**

com•put•ing /kəm'pyo͞otiɴɢ/ ▸ **n.** the use or operation of computers:

developments in mathematics and computing | [as adjective] *computing facilities.*

con•cat•e•nate /kən'kætn,āt/ ▸ **v.** [trans.] link (character strings, files, commands, etc.) together in a chain or series to save space or handle them as a unit.

ORIGIN late 15th cent. (as an adjective): from late Latin *concatenat-* 'linked together,' from the verb *concatenare*, from *con-* 'together' + *catenare*, from *catena* 'chain.'

con•cat•e•na•tion /kən,kætn'āSHən/ ▸ **n.** a series of connected character strings, files, etc.

■ the action of linking character strings, files, etc., together in a series. ■ the condition of being linked in such a way.

con•fig•u•ra•tion /kən,figyə'rāSHən/ ▸ **n.** the arrangement in which items of computer hardware or software are interconnected: *it comes with a removable hard disk drive as part of the standard configuration.*

ORIGIN mid 16th cent. (denoting the relative position of celestial objects): from late Latin *configuratio(n-)*, from Latin *configurare* 'shape after a pattern' (see **CONFIGURE**).

con•fig•ure /kən'figyər/ ▸ **v.** [trans.] (often **be configured**) include as part of (a computer system) particular elements and hardware devices.

■ set up (computer hardware or software) to carry out a designated task: *expanded memory can be configured as a virtual drive* | *you can configure how you want virus detections handled.*

DERIVATIVES **con•fig•ur•a•ble adj.**

ORIGIN late Middle English: from Latin *configurare* 'shape after a pattern,' from *con-* 'together' + *figurare* 'to shape' (from *figura* 'shape or figure').

con•nec•tion•ism /kə'nekSHə,nizəm/ ▸ **n.** an artificial intelligence approach to cognition in which multiple connections between nodes (equivalent to brain cells) form a massive interactive network in which many processes take place simultaneously. Certain processes in this network, operating in parallel, are grouped together in hierarchies that bring about results such as thought or action (also known as **parallel distributed processing**).

con•nec•tiv•i•ty /kə,nek'tivitē/ ▸ **n.** capacity for the interconnection of platforms, systems, and applications: *connectivity between Sun and Mac platforms.*

con•tent pro•vid•er /'käntent prə,vīdər/ ▸ **n.** a person or organization who supplies information for use on a Web site: *the content provider for short law and practice news updates* | *he worked for an Internet content provider.*

con•trol /kən'trōl/ ▸ **n.** short for **CONTROL KEY**.

con•trol char•ac•ter /kən'trōl ˌkæriktər/ ▸ **n.** a character that does not represent a printable character but serves to initiate a particular action.

con•trol key /kən'trōl ˌkē/ ▸ **n.** a key that alters the function of another key if both are pressed at the same time.

con•vert•er /kən'vərtər/ (also **con•ver•tor**) ▸ **n.** a program that converts data from one format to another.

cook•ie /'ko͝okē/ ▸ **n.** (pl. **cook•ies**) a packet of data sent by an Internet server to a browser, which is returned by the browser each time it subsequently accesses the same server, used to identify the user or track their access to the server.

ORIGIN early 18th cent.: from Dutch *koekje* 'little cake,' diminutive of *koek*. The original form for the computing sense was *magic cookie*, used by Unix programmers.

co•proc•es•sor /'kō,präsesər/ ▸ **n.** a processor designed to supplement the capabilities of the primary processor.

cop•y /'käpē/ ▸ **v.** (**cop•ies, cop•ied**) [trans.] reproduce (data stored in one location) in another location: *the command will* ***copy*** *a file* ***from*** *one disc* ***to*** *another.*

ORIGIN Middle English (denoting a transcript or copy of a document): from Old French *copie* (noun), *copier* (verb), from Latin *copia* 'abundance' (in medieval Latin 'transcript,' from such phrases as *copiam describendi facere* 'give permission to transcribe').

core /kôr/ ▸ **n.** a tiny ring of magnetic material formerly used in a computer memory to store one bit of data, now superseded by semiconductor memory chips.

ORIGIN Middle English: of unknown origin.

cor•rupt /kə'rəpt/ ▸ **adj.** (of a computer database or program) having errors introduced.

▸**v.** [trans.] (often **be corrupted**) cause errors to appear in (a computer program or database): *a program that has somehow corrupted your system files.*

DERIVATIVES **cor•rupt•i•ble adj.**

ORIGIN Middle English: from Latin *corruptus*, past participle of *corrumpere* 'mar, bribe, destroy,' from *cor-* 'altogether' + *rumpere* 'to break.'

cor•rup•tion /kə'rəpsHən/ ▸ **n.** the process of causing errors to appear in a computer program or database.

ORIGIN Middle English: via Old French from Latin *corruptio(n-)*, from *corrumpere* 'mar, bribe, destroy' (see **CORRUPT**).

cps (also **c.p.s.**) ▸ **abbr.** characters per second.

CPU ▸ **abbr.** central processing unit, the part of a computer in which operations are controlled and executed. Also see **PROCESSOR**.

■ a computer case containing the processor, drives, audio and video circuit boards, peripheral connectors, etc.: [as adj.] *a computer desk with a CPU shelf.*

crash /kræsH/ ▸ **v.** [intrans.] (of a machine, system, or software) fail suddenly: *the project was postponed because the computer crashed.*

▸**n.** a sudden failure that puts a system out of action.

crawl•er /'krôlər/ ▸ **n.** a program that searches the World Wide Web, typically in order to create an index of data.

CRC ▸ **abbr.** cyclic redundancy check, a data code that detects errors during transmission, storage, or retrieval.

crip•ple•ware /'kripəl,we(ə)r/ ▸ **n.** informal software distributed with reduced functionality with a view to attracting payment for a fully functional version: *if the software blocks crucial functions such as saving files, it's sneered at as "crippleware," but if it's only missing high-end options, it may provide everything you need.*

cross-as•sem•bler /'krôs ə,semblər/ ▸ **n.** an assembler that can convert instructions into machine code for a computer other than that on which it is run.

cross-com•pil•er /'krôs kəm,pīlər/ ▸ **n.** a compiler that can convert instructions into machine code or low level code for a computer other than that on which it is run.

cross-plat•form /'krôs 'plæt,fôrm/ ▸ **adj.** able to be used on different types of computers or with different software packages: *a cross-platform game.*

cross-post /ˈkrôs ˌpōst/ (also **cross•post**) ▸ **v.** post a single message to multiple Internet newsgroups or readings lists: [intrans.] *if you cross-post to both lists, several hundred people will receive two copies of your message.*

■ repost a message appearing on one list or newsgroup to another: [trans.] *please do not cross-post this vacancy.*

▸**n.** a message posted to more than one newsgroup or reading list: *please note that while we do not have any explicit policy about crossposting to other newsgroups, we do strongly recommend that the crossposts be to relevent newsgroups only.*

cross-post•ing /ˈkrôs ˈpōstiNG/ ▸ **n.** the simultaneous sending of a message to more than one newsgroup or other distribution system on the Internet in such a way that the receiving software at individual sites can detect and ignore duplicates.

CRT ▸ **abbr.** cathode ray tube, a high-vacuum tube in which cathode rays produce a luminous image on a fluorescent screen, used chiefly in televisions and computer monitors.

■ a computer monitor having such a tube.

cryp•to /ˈkriptō/ ▸ **n.** short for **CRYPTOGRAPHY**.

C2C ▸ **abbr.** consumer-to-consumer, denoting transactions conducted via the Internet between consumers.

cur•sor /ˈkərsər/ ▸ **n.** a movable indicator on a computer screen identifying the point that will be affected by input from the user, for example showing where typed text will be inserted.

ORIGIN Middle English (denoting a runner or running messenger): from Latin, 'runner,' from *curs-*.

cut /kət/ ▸ **v.** (**cut•ting**; past and past part. **cut**) [trans.] delete (part of a text or other display) completely or so as to insert a copy of it elsewhere. See also **CUT AND PASTE**.

cut and paste /ˈkət ən(d) ˈpāst/ ▸ **n.** a process used in assembling text on a computer, in which items are removed from one part and inserted elsewhere.

▸ **v.** [trans.] move (an item of text) using this technique.

cut•scene /ˈkətˌsēn/ ▸ **n.** (in computer games) a scene that develops the storyline and is often shown on completion of a certain level, or when the player's character dies.

cy•ber•at•tack /ˈsībərəˌtak/ ▸ **n.** an effort by hackers to damage or destroy a computer network or system: *warning that America is increasingly vulnerable to cyberattacks, President Clinton ordered the strengthening of defenses against terrorists, germ warfare and other unconventional security threats of the 21st century.*

DERIVATIVES **cy•ber•at•tack•er n.**

cy•ber•cash /ˈsībərˌkasʜ/ (also **CyberCash**) ▸ **n. 1** funds used in electronic financial transactions, especially over the Internet: *there's a form of cybercash that cannot be linked to an owner or spender.*

2 money stored on an electronic smart card or in an online credit account: *but we can have something delivered over the Net, as long as we've got a walletful of CyberCash.*

cy•ber•crime /ˈsībərˌkrīm/ ▸ **n.** crime conducted via the Internet or some other computer network: *in terms of lost sales and productivity, the bureau estimates cybercrime costs Americans more than $10 billion a year.*

cy•ber•law /ˈsībərˌlô/ ▸ **n.** laws, or a specific law, relating to Internet and computer offenses, especially fraud, copyright infringement, etc.: *cyberlaw encompasses the regulations that govern cyberspace, and it's such a cutting-edge field that the search of the legal database Westlaw brings up no federal cases using the term.*

cy•ber•mall /ˈsībərˌmôl/ ▸ **n.** a commercial Web site through which a range of goods may be purchased; a virtual shopping mall on the Internet: *a typical example would be shopping in a cybermall, where a user might move from store to store and acquire various merchandise along the way.*

cy•ber•naut /ˈsībərˌnôt; -ˌnät/ ▸ **n.** a person who wears sensory devices in order to experience virtual reality: *a "cybernaut" can step through walls and wander at will.*

■ a person who uses the Internet.

ORIGIN 1990s: from *cyber-*, on the pattern of *astronaut* and *aeronaut.*

cy•ber•pet /ˈsībərˌpet/ ▸ **n.** an electronic toy that simulates a real pet and with which human interaction is possible. (Also called **dig•i•pet** or **vir•tu•al pet**.)

cy•ber•pho•bi•a /ˌsībər'fōbēə/ ▸ **n.** fear of or anxiety about computing or information technology; reluctance to use computers, especially the Internet.

DERIVATIVES **cy•ber•phobe** /'sībərˌfōb/ **n.**; **cy•ber•pho•bic adj.**

cy•ber•porn /'sībərˌpôrn/ ▸ **n.** pornography viewable on a computer screen, especially accessed on the Internet: *my topic is cyberporn, that is, computerized pornography.*

cy•ber•punk /'sībərˌpəNGk/ ▸ **n.** a genre of science fiction set in a transgressive subculture of an oppressive society dominated by computer technology: *he has been at the forefront of the cyberpunk* | [as adjective] *futuristic cyberpunk fantasies.*

■ a writer of such science fiction. ■ a person who accesses computer networks illegally, esp. with malicious intent.

cy•ber•shop /'sībərˌSHäp/ ▸ **v.** [intrans.] (**cy•ber•shopped, cy•ber•shop•ping**) [often as noun] (**cybershopping**) purchase or shop for goods and services on a Web site: *research commissioned by Elron Software found that more than half of American workers cybershop on company time.*

▸**n.** (also **cy•ber•store** /'sībərˌstôr/) a Web site that sells or provides information about retail goods or services: *the retailer's cybershop sometimes has different prices than in its mail-order catalog.*

DERIVATIVES **cy•ber•shop•per n.**

cy•ber•slack•er /'sībərˌslækər/ ▸ **n.** informal a person who uses their employer's Internet and e-mail facilities for personal activities during working hours: *cyberslackers who download and distribute pornography waste productive time, consume expensive bandwidth and could be in breach of sexual harassment laws.*

DERIVATIVES **cy•ber•slack•ing n.**

cy•ber•space /'sībərˌspās/ ▸ **n.** the notional environment in which communication over computer networks occurs.

cy•ber•squat•ting /'sībərˌskwätiNG/ ▸ **n.** the practice of registering names, especially well-known company or brand names, as Internet domains, in the hope of reselling them at a profit: *a United Nations panel late last week issued a set of proposed rules that seek to* *end so-called cybersquatting—registering popular corporate or product names as Internet addresses in order to sell them back to the trademark holders or to divert Internet traffic from those companies.*

DERIVATIVES **cy•ber•squat•ter n.**

cy•ber•stalk•ing /ˈsībərˌstôkiNG/ ▸ **n.** the repeated use of electronic communications to harass or frighten someone, for example by sending threatening e-mails: *fewer than one-third of states currently have anti-stalking laws that explicitly cover cyberstalking.*

DERIVATIVES **cy•ber•stalk•er n.**

cy•ber•surf•er /ˈsībərˌsərfər/ ▸ **n.** a person who habitually uses or browses the Internet.

DERIVATIVES **cy•ber•surf•ing n.**

cy•ber•ter•ror•ism /ˌsībərˈterəˌrizəm/ ▸ **n.** the politically motivated use of computers and information technology to cause severe disruption or widespread fear in society: *a spending bill to finance the Justice Department that would make it easier for law enforcement to wiretap computers and combat cyberterrorism.*

DERIVATIVES **cy•ber•ter•ror•ist n.**

cy•ber•war /ˈsībərˌwôr/ ▸ **n.** acts of hostility carried out on the Internet against national interests or ethnic groups: *but this exploit took cyberwar a step further: the attacker stole some 3,500 e-mail addresses and 700 credit card numbers, sent anti-Israeli diatribes to the addresses, and published the credit card data on the Internet.*

cy•brar•i•an /sīˈbre(ə)rēən/ ▸ **n.** a librarian or researcher who uses the Internet as an information resource: *find out what cybrarians involved in electronic information services think about these and other questions related to libraries and the national information infrastructure.*

ORIGIN blend of *cyber-* and *librarian.*

cy•cle /ˈsīkəl/ ▸ **n.**

▸ **v.** [intrans.] move in or follow a regularly repeated sequence of operations.

ORIGIN late Middle English: from Old French, from late Latin *cyclus*, from Greek *kuklos* 'circle.'

cy•clic re•dun•dan•cy check /ˈsīklik riˈdəndənsē ˌCHek/ (also **cy•clic re•dun•dan•cy code**) (abbr.: **CRC**) ▸ **n.** a code added to data, used to detect errors occurring during transmission, storage, or retrieval.

cy•pher•punk /ˈsīfərˌpəNGk/ ▸ **n.** a person who uses encryption when accessing a computer network in order to ensure privacy, esp. from government authorities.

ORIGIN 1990s: on the pattern of *cyberpunk*.

D

dae•mon /ˈdēmən/ (also **de•mon**) ▸ **n.** a background process that handles requests for services such as print spooling and file transfers, and is dormant when not required: *a background-only application (daemon) that allows you to control the resolutions of all your displays | the daemon notes the IP where the e-mail originated.*

ORIGIN 1980s: perhaps from *d(isk) a(nd) e(xecution) mon(itor)* or from *de(vice) mon(itor)*, or merely a transferred use of **DEMON**.

dai•sy chain /ˈdāzē ˌCHān/ (also **dai•sy-chain**) ▸ **v.** [trans.] connect (several devices) together in a linear series: *the keyboard has a USB port so you can daisy chain a mouse, joystick, or other device to it.*

DERIVATIVES **dai•sy-chain•a•ble adj.**

dai•sy wheel /ˈdāzē ˌ(h)wēl/ ▸ **n.** a device used as a printer in word processors and typewriters, consisting of a disk of spokes extending radially from a central hub, each terminating in a printing character. [as adj] *a daisy-wheel printer.*

da•ta /ˈdæṯə; ˈdāṯə/ ▸ **n.** the quantities, characters, or symbols on which operations are performed by a computer, being stored and transmitted in the form of electrical signals and recorded on magnetic, optical, or mechanical recording media.

■ facts and statistics collected together for reference or analysis.

ORIGIN mid 17th cent. (as a term in philosophy): from Latin.

USAGE: **Data** was originally the plural of the Latin word *datum*, 'something (e.g., a piece of information) given.' **Data** is now used as a singular where it means 'information': *this data was prepared for the conference.* It is used as a plural in technical contexts and when the collecion of bits of information is stressed: *the data are* *stored on the hard drive.* Avoid *datas* and *datae*, which are false plurals, neither English nor Latin.

da•ta bank /ˈdætə ˌbæNGk; ˈdātə/ (also **da•ta•bank** /ˈdætəˌbæNGk; ˈdā-/) ▸ **n.** a large repository of data on a particular topic, sometimes formed from more than one database, and accessible by many users.

da•ta•base /ˈdætəˌbās; ˈdātə/ ▸ **n.** a structured set of data held in a computer, esp. one that is accessible in various ways: *a comprehensive attorney database.*

da•ta•base man•age•ment sys•tem /ˈdætəˌbās ˌmanijmənt ˌsistəm; ˈdātə-/ (abbr.: **DBMS**) ▸ **n.** software that handles the storage, retrieval, and updating of data in a computer system.

da•ta dic•tion•ar•y /ˈdætə ˌdiksHənerē; ˈdātə/ ▸ **n.** a set of information describing the contents, format, and structure of a database and the relationship between its elements, used to control access to and manipulation of the database.

da•ta•glove /ˈdætəˌgləv; ˈdātə-/ ▸ **n.** a device, worn like a glove, that allows the manual manipulation of images in virtual reality.

da•ta•link /ˈdætəˌliNGk; ˈdātə-/ ▸ **n.** an electronic connection for the exchange of information: *NASA is working on a datalink system that would allow aircraft controllers and pilots to exchange electronic messages that could be viewed on video displays.*

da•ta min•ing /ˈdætə ˌmīniNG; ˈdātə/ ▸ **n.** the practice of examining large databases in order to generate new information: *the committee is holding hearings on issues of privacy and data mining.*

da•ta•point /ˈdætəˌpoint; ˈdātə-/ ▸ **n.** an identifiable element in a data set: *software that can quickly process tens of thousands of datapoints.*

da•ta set /ˈdætə ˌset; ˈdātə/ ▸ **n.** a collection of related sets of information that is composed of separate elements but can be manipulated as a unit by a computer.

da•ta smog /ˈdætə ˌsmäg; ˈdātə/ ▸ **n.** informal an overwhelming excess of information, especially from the Internet: *nowadays, people need help getting their intellectual bearings because cable has become a torrent of ideology, dueling experts, and data smog.*

da•ta type /ˈdætə ˌtīp; ˈdātə/ ▸ **n.** a particular kind of data item, as defined by the values it can take, the programming language used, or the operations that can be performed on it.

da•ta ware•house /ˈdæt̲ə ˈwe(ə)rˌhows; ˈdāt̲ə/ ▸ **n.** a large store of data accumulated from a wide range of sources within a company and used to guide management decisions.

DERIVATIVES **da•ta ware•hous•ing n.**

daugh•ter•card /ˈdôtərˌkärd/ (also **daugh•ter•board** /ˈdôtərˌbôrd/) ▸ **n.** an expansion circuit card affixed to a motherboard that accesses memory and the CPU directly rather than through a bus.

DBMS ▸ **abbr.** database management system.

DDE ▸ **n.** a standard allowing data to be shared between different programs.

ORIGIN 1980s: abbreviation of *Dynamic Data Exchange*.

DDoS (also **DDOS**) ▸ **abbr.** distributed denial of service, the intentional paralyzing of a computer network by flooding it with data sent simultaneously from many individual computers: [as modifier] *automation of so-called zombies, which are then used to stage DDOS attacks, has made it a lot easier to launch an attack from a single terminal.*

DDR ▸ **abbr.** double data rate, a technology that allows computer memory to send twice the amount of data to the processor, increasing the speed of the computer.

de•bug /dēˈbəg/ ▸ **v.** (**de•bugged**, **de•bug•ging**) [trans.] identify and remove errors from (computer hardware or software): *games are the worst to debug* | [as noun] (**debugging**) *software debugging.*

▸ **n.** the process of identifying and removing errors from computer hardware or software.

de•bug•ger /dēˈbəgər/ ▸ **n.** a computer program that assists in the detection and correction of errors in computer programs.

de•com•pile /ˌdēkəmˈpīl/ ▸ **v.** [trans.] produce source code from (compiled code).

DERIVATIVES **de•com•pi•la•tion** /dēˌkämpəˈlāsʜən; ˌdēkäm-/ **n.**; **de•com•pil•er n.**

de•com•press /ˌdēkəmˈpres/ ▸ **v.** [trans.] expand (compressed computer data) to its normal size so that it can be read and processed by a computer.

DERIVATIVES **de•com•pres•sion** /-ˈpresʜən/ **n.**

de•crypt /dē'kript/ ▸ v. [trans.] convert (encrypted data) into an intelligible form.

DERIVATIVES **de•cryp•tion** /-'kripSHən/ n.

ORIGIN 1930s: from *de-* (expressing reversal) + *crypt* as in *encrypt.*

ded•i•cat•ed /'dedi,kātid/ ▸ adj. (of a thing) exclusively allocated to or intended for a particular service or purpose: *a dedicated telephone line.*

deep mag•ic /'dēp 'majik/ ▸ n. any of the techniques used in the development of software or computer systems that require the programmer to have esoteric theoretical knowledge: *some hackers use deep magic to tweak existing software.*

ORIGIN probably from C.S. Lewis's *Narnia* books.

de•fault /di'fôlt/ ▸ n. a preselected option adopted by a computer program or other mechanism when no alternative is specified by the user or programmer: *the default font is Times Roman* | [as adjective] *default settings.*

▸ v. [intrans.] (**default to**) (of a computer program or other mechanism) revert automatically to (a preselected option): *when you start a fresh letter, the system will default to its own style.*

de•frag ▸ v. /dē'fræg/ [trans.] (**de•fragged**, **de•frag•ging**) defragment files (on a disk): *we defragged the disk drives and did some more tweaking.*

▸n. /'dē,fræg/ an instance of defragmenting a disk, or the utility that does this: *maybe you leave it running to collect e-mails or carry out a defrag.*

ORIGIN shortened from *defragment.*

de•frag•ment /dē'fræg,ment/ ▸ v. [trans.] reduce the fragmentation of (a disk) using a software program that concatenates pieces of files stored in separate locations by rewriting them to contiguous sectors: *the safe way to defragment your files.*

DERIVATIVES **de•frag•men•ta•tion** /dē,frægmən'tāSHən/ n.; **de•frag•ment•er** n.

de•in•stall /,dē-in'stôl/ (*British* also **de•in•stal**) ▸ v. (**de•in•stalls** also **de•in•stals**; **de•in•stalled**; **de•in•stall•ing**) [trans.] remove (an application or file) from a computer; uninstall.

DERIVATIVES **de•in•stal•la•tion** /,dē-instə'lāSHən/ n.; **de•in•stall•er** n.

de•lete /di'lēt/ ▸ **v.** [trans.] (usu. **be deleted**) remove (data) from a computer's memory: *delete the file | delete the last three lines.*

■ remove or obliterate (written matter) by typing a line through it: *the passage was deleted.*

▸ **n.** a command or key on a computer that erases text: *highlight the text and press delete.*

ORIGIN late Middle English: from Latin *delet-* 'blotted out, effaced,' from the verb *delere.*

dem•o /'demō/ informal ▸ **n.** (pl. **dem•os**) a demonstration of the capabilities of something, typically new hardware or software: [as adjective] *a demo disk.*

▸ **v.** (**dem•os, dem•oed**) [trans.] demonstrate the capabilities of (software or equipment).

de•mon /'dēmən/ ▸ **n.** variant spelling of **DAEMON**.

de•ni•al-of-serv•ice at•tack /di'nīəl əv 'sərvis ə,tæk/ ▸ **n.** an overwhelmingly large amount of data sent to a network in an intentional effort to disrupt its operation. Compare with **MAIL BOMB**.

de•queue /dē'kyōō/ ▸ **v.** [trans.] (**de•queued, de•queu•ing** or **de•queue•ing**) remove (an item) from a queue.

DES ▸ **abbr.** data encryption standard.

de•scrip•tor /di'skriptər/ ▸ **n.** a piece of stored data that indicates how other data is stored.

de•se•lect /,dēsə'lekt/ ▸ **v.** [trans.] turn off (a selected feature) on a list of options on a computer menu.

■ remove highlighting from a selected item in a list or from text on a computer screen.

DERIVATIVES **de•se•lec•tion n.**

desk•top /'desk,täp/ ▸ **n.** a computer that is suitable for use at an ordinary desk but is too large to carry with you: [as adj] *a desktop PC.* Compare with **NOTEBOOK**.

■ the working area of a computer screen regarded as a representation of a notional desktop and containing icons representing items such as files and a wastebasket.

desk•top pub•lish•ing /'desk,täp 'pəbliSHiNG/ (abbr.: **DTP**) ▸ **n.** the production of printed matter by means of a printer linked to a desktop computer, with special software. The system enables reports, advertising matter, company magazines, etc., to be produced cheaply with a layout and print quality similar to that of typeset books, for xerographic or other reproduction.

de•vice /di'vīs/ ▸ **n.** a piece of equipment designed to be connected to a computer, such as a printer, network card, camera dock, etc.

DHTML ▸ **abbr.** dynamic HTML, a collection of web browser enhancements that enable dynamic and interactive features on web pages.

di•ag•nos•tic /ˌdīəg'nästik/ ▸ **n.** a program or routine that helps a user to identify errors.

ORIGIN early 17th cent.: from Greek *diagnōstikos* 'able to distinguish,' from *diagignōskein* 'distinguish'; the noun from *hē diagnōstikē tekhnē* 'the art of distinguishing.'

di•al•er /'dī(ə)lər/ ▸ **n.** a device or piece of software for calling telephone numbers automatically: *hackers can break in with speed dialers.*

di•al-in /'dī(ə)l ˌin/ ▸ **adj.** another term for **DIAL-UP**.

di•a•log box /'dīəˌlôg ˌbäks/ ▸ **n.** a small area on screen, in which the user is prompted to provide information or select commands: *the application presents you with a dialog box to choose how to save your file.*

di•al-up /'dī(ə)l ˌəp/ ▸ **adj.** (of a computer system or service) used remotely via a telephone line: *dial-up Internet access | a dial-up connection.*

dic•tion•ar•y at•tack /'dikSHəˌnerē əˌtæk/ ▸ **n.** an attempted illegal entry to a computer system that uses a dictionary headword list to generate possible passwords: *it's going to be pretty hard to argue that someone doing a dictionary attack on a hotmail or an AOL or a Yahoo has actually been using personal information.*

diff /dif/ ▸ **v.** [trans.] compare (files) electronically in order to determine how or whether they differ: *I diffed my new XF86config file with the old one and found the only diffs were that I had uncommented the lines referring to linear addressing.*

▸ **n.** such an electronic comparison.

di•ge•ra•ti /dijə'rätē/ ▸ **plural n.** people with expertise or professional involvement in information technology.

ORIGIN 1990s: blend of **DIGITAL** and *literati.*

dig•i•cam /'dijiˌkam/ ▸ **n.** a digital camera.

dig•i•pet /ˈdijiˌpet/ ▸ **n.** see **CYBERPET.**

dig•it•al /ˈdijitl/ ▸ **adj.** relating to or using signals or information represented by discrete values (digits) of a physical quantity, such as voltage or magnetic polarization, to represent arithmetic numbers or approximations to numbers from a continuum or logical expressions and variables: *a digital monitor.* Often contrasted with **ANALOG.**

DERIVATIVES **dig•it•al•ly adv.**

dig•it•al cam•er•a /ˈdijitl ˈkam(ə)rə/ ▸ **n.** a camera that records and stores digital images.

dig•it•al di•vide /ˈdijitl diˈvīd/ ▸ **n.** the gulf between those who have ready access to computers and the Internet, and those who do not: *these results, if confirmed by other studies, point to a worrying "digital divide" based on race, gender, educational attainment and income.*

dig•it•al sig•na•ture /ˈdijitl ˈsignəCHər; -ˌCHo͝or/ ▸ **n.** a digital code (generated and authenticated by public key encryption) that is attached to an electronically transmitted document to verify its contents and the sender's identity.

dig•i•tal sub•scrib•er line /ˈdijitl səbˈskrībər ˌlīn/ ▸ **n.** see **DSL**

digital-to-analog converter /ˈdijitl to͞o ˈænlˌôg kənˌvərtər/ ▸ **n.** an electronic device for converting digital signals to analog form.

dig•i•tal ver•sa•tile disc /ˈdijitl ˈvərsətl ˈdisk/ ▸ **n.** see **DVD**.

dig•i•tal vid•e•o re•cord•er /ˈdijitl ˈvidēō riˌkôrdər/ ▸ **n.** (abbreviation **DVR**) a programmable electronic device that writes audio and video input, typically from a television signal, to a rewritable hard disk.

dig•i•ta•tion /ˌdijiˈtāSHən/ ▸ **n.** the process of converting data to digital form.

dig•i•tize /ˈdijiˌtīz/ ▸ **v.** [trans.] [usu. as adjective] (**digitized**) convert (pictures or sound) into a digital form that can be processed by a computer.

DERIVATIVES **dig•i•ti•za•tion** /ˌdijitəˈzāSHən/ **n.**; **dig•i•tiz•er n.**

DIMM ▸ **abbr.** dual in-line memory module, containing RAM chips and having a 64-bit data path to the computer. Compare with **SIMM**.

DIP /dip/ ▸ **abbr.** ■ document image processing, a system for the digital storage and retrieval of documents as scanned images. ■ dual in-line package, a package for an integrated circuit consisting of a rectangular sealed unit with two parallel rows of downward-pointing pins.

DIP switch /ˈdip ˌswiCH/ ▸ **n.** an arrangement of switches in a dual in-line package used to select the operating mode of a device such as a printer.

di•rect ac•cess /dəˈrekt ˈækses/ ▸ **n.** the facility of retrieving data immediately from any part of a computer file, without having to read the file from the beginning. Compare with **RANDOM ACCESS** and **SEQUENTIAL ACCESS.**

di•rec•to•ry /dəˈrekt(ə)rē/ ▸ **n.** (pl. **di•rec•to•ries**) a file, often represented by a folder on a computer screen, that consists solely of a set of other files (which may themselves be directories).

ORIGIN late Middle English: from late Latin *directorium*, from *director* 'governor,' from *dirigere* 'to guide.'

dis•as•sem•ble /ˌdisəˈsembəl/ ▸ **v.** [trans.] (often **be disassembled**) translate (a program) from machine code into a symbolic language.

disc /disk/ ▸ **n.** variant spelling of **DISK.**

dis•cus•sion board /diˈskəSHən ˌbôrd/ ▸ **n.** another term for **MESSAGE BOARD.**

disk /disk/ (also **disc**) ▸ **n.** an information storage device for a computer in the shape of a round flat plate that can be rotated to give access to all parts of the surface. The data may be stored either magnetically (in a **mag•net•ic disk** such as a hard disk) or optically (in an **op•ti•cal disk** such as a CD-ROM).

■ (**disc**) short for **COMPACT DISC.** ■ (**disc**) a **DVD**: *the collector's DVD set includes a 20-page book and bonus disc.*

DERIVATIVES **disk•less adj.**

ORIGIN mid 17th cent. (originally referring to the seemingly flat round form of the sun or moon): from French *disque* or Latin *discus*.

USAGE: Generally speaking, the U.S. spelling is **disk** and the British spelling is **disc**, although there is much overlap and variation between the two. In particular the spelling for senses relating to computers is nearly always **disk**, as in **floppy disk**, **disk drive**, etc., but **compact disc**.

disk drive /ˈdisk ˈdrīv/ ▸ **n.** a device that allows a computer to read from and write to computer disks.

disk•ette /diˈsket/ ▸ **n.** another term for a **FLOPPY DISK**.

disk op•er•at•ing sys•tem /ˈdisk ˌäpəˌrātiNG ˌsistəm/ ▸ **n.** see **DOS**.

dis•mount /disˈmownt/ ▸ **v.** [trans.] make (a disk or disk drive) unavailable for use.

dis•play /diˈsplā/ ▸ **v.** [trans.] (of a computer or other device) show (information) on a screen.

▸ **n.** an electronic device for the visual presentation of data: *a 17-inch color display* | [as adjective] *a visual display screen.*

■ the process or facility of presenting data on a computer screen or other device: *the processing and display of high volumes of information.* ■ the data shown on a computer screen or other device.

dis•played /diˈsplād/ ▸ **adj.** (of information) shown on a computer screen or other device: *a utility designed to allow you to cut up pieces of displayed graphics.*

dis•trib•ut•ed /diˈstribyo͞otid/ ▸ **adj.** (of a computer system) spread over several machines, especially over a network: *the Wi-Fi IDS would package two new features from IBM Research, including the capability to do security auditing in a distributed environment.*

dis•trib•ut•ed sys•tem /disˈtribyo͞otid ˌsistəm/ ▸ **n.** a number of independent computers linked by a network.

dis•tro /ˈdistrō/ ▸ **n.** a distributor, especially of Linux software or of web-based zines: *I've been working on this project for a little while and decided to post this here in case anyone who runs a distro is interested.*

■ a particular distributable or distributed version of Linux software: *I was excited enough about this distro that I forked over the cash to buy it.*

dith•er /ˈdiTHər/ ▸ **v.** [trans. & intrans.] display or print (an image) without sharp edges so that there appear to be more colors in it than are really available: [as adjective] (**dithered**) *dithered bit maps.*

DERIVATIVES **dith•er•er n.**

DLL ▸ **abbr.** dynamic link library, a collection of small programs for common use by larger programs or software suites.

■ a particular file containing such a program, and carrying the extension .DLL: *a program that has been corrupted by a DLL.*

DMA ▸ **abbr.** direct memory access, a method allowing a peripheral device to transfer data to or from the memory of a computer system using operations not under the control of the central processor: *DMA can lower power consumption and increase speed.*

DMCA ▸ **abbr.** the Digital Millennium Copyright Act, a 1998 U.S. law that was intended to update copyright law for electronic commerce and electronic content providers. It criminalizes the circumvention of electronic and digital copyright protection systems: *this is a classic fair-use right protected under the copyright act, but it is a right that is extinguishable under the DMCA.*

DNS ▸ **abbr.** domain name system, the system by which Internet domain names and addresses are tracked and regulated.

doc /däk/ ▸ **abbr.** document.

dock /däk/ ▸ **n.** a DOCKING STATION.

▸ **v.** [intrans.] attach (a piece of equipment) to another: *the user wants to dock a portable into a desktop computer.*

ORIGIN late Middle English: from Middle Dutch, Middle Low German *docke*, of unknown origin.

dock•ing sta•tion /ˈdäkiNG ˌstāSHən/ ▸ **n.** a device to which a portable computer is connected so that it can be used like a desktop computer, with an external power supply, monitor, data transfer capability, etc.: *for use at a desk, you snap the tablet PC into a docking station.*

■ a device with which a piece of equipment is attached to a computer to allow transfer of data: *the camera has a docking station for transferring images to your computer.*

doc•u•men•ta•tion /ˌdäkyəmenˈtāSHən/ ▸ **n.** the written specification and instructions accompanying a computer program or hardware: *user-friendly documentation.*

DOI ▸ **abbr.** digital object identifier, a unique identifying number allocated to a Web site.

do•main /dōˈmān/ ▸ **n.** a distinct subset of the Internet with addresses sharing a common suffix, such as the part in a particular country or used by a particular group of users: *country-specific Internet domains such as .ca (Canada)* [as adj] *to have a Web site, you need a domain name.*

ORIGIN late Middle English (denoting heritable or landed property): from French *domaine*, alteration (by association with Latin *dominus* 'lord') of Old French *demeine* 'belonging to a lord'.

do•main name /dō'mān ˌnām/ ▸ **n.** a series of alphanumeric strings separated by periods, such as *www.oup.com*, serving as an address for a computer network connection and identifying the owner of the address. The last three letters in a domain name indicate what type of organization owns the address: for instance, .com stands for commercial, .edu for educational, and .org for nonprofit.

don•gle /'däNGgəl/ ▸ **n.** an electronic device that must be attached to a computer in order to use protected computer software.

ORIGIN 1980s: an arbitrary formation.

DoS ▸ **abbr.** denial of service, an interruption in an authorized user's access to a computer network, typically one caused with malicious intent.

DOS /dôs/ ▸ **abbr.** disk operating system, an operating system originally developed for IBM personal computers.

dot-bomb /'dät 'bäm/ (also **dot bomb** or **dot.bomb**) ▸ **n.** informal an unsuccessful dot-com: *many promising Internet start-ups ended up as dot-bombs.*

DERIVATIVES **dot-bomb v.** [intrans.]

dot ma•trix /'dät 'mātriks/ ▸ **n.** [usu. as adjective] a grid of dots that are filled selectively to produce an image on a screen or paper: *a dot matrix display board.*

dot ma•trix print•er /'dät 'mātriks ˌprintər/ ▸ **n.** a printer that forms images of letters, numbers, etc., from a number of tiny dots.

dou•ble-click /'dəbəl 'klik/ ▸ **v.** [intrans.] press a mouse button twice in quick succession to select a file, program, or function: *to run a window just **double-click on** the icon.*

■ [trans.] select (a file) in this way.

down /down/ ▸ **adv.** (of a computer system) out of action or unavailable for use (esp. temporarily): *the system went down yesterday.*

▸ **adj.** [predic.] (of a computer system) temporarily out of action or unavailable: *sorry, but the computer's down.*

ORIGIN Old English *dūn*, *dūne*, shortened from *adūne* 'downward,' from the phrase *of dūne* 'off the hill'.

down•code /ˈdownˌkōd/ ▸ **v.** rewrite or convert (programs or software) into a lower level language: [trans.] *some of the libraries written into C were also downcoded into assembly primarily for compactness but with the side effect of speeding them up.*
DERIVATIVES **down•cod•ing n.**

down•lev•el /ˈdownˌlevəl/ ▸ **adj.** using an earlier version of software, hardware, or an operating system: *even if all the vulnerabilities were fixed tomorrow morning in all of the products, there's still 600 million computers, many of them downlevel, many of them on funny versions that wouldn't have all of these vulnerabilities patched, fixed, and up to date.*

down•load /ˈdownˌlōd/ ▸ **v.** [trans.] copy (data) from one computer system to another or to a disk: *purchase and download music.*
▸ **n.** the act or process of copying data in such a way: [as adjective] *a download and upload routine.*
■ a computer file transferred in such a way: *a popular download from bulletin boards.*
DERIVATIVES **down•load•a•ble adj.**

down•time /ˈdownˌtīm/ (also **down time**) ▸ **n.** time during which a computer is out of action or unavailable for use.

DP ▸ **abbr.** data processing.

dpi ▸ **abbr.** dots per inch, a measure of the resolution of printers, scanners, etc.

draft /dræft/ ▸ **n.** (in full **draft mode** /ˈdræft ˌmōd/) a mode of operation of a printer in which text is produced rapidly but with relatively low definition.

drag /dræg/ ▸ **v.** (**dragged, drag•ging**) [trans.] move (an icon or other image) across a computer screen using a tool such as a mouse: *you simply drag the file to the new folder.*
ORIGIN Middle English: from Old English *dragan* or Old Norse *draga* 'to draw'; the noun partly from Middle Low German *dragge* 'grapnel.'

drag-and-drop /ˈdræg ən ˈdräp/ ▸ **v.** [trans.] move (an icon or other screen element) to another part of the screen using a mouse or similar device, typically in order to perform some operation on a file or document.

▸ **adj.** of, relating to, or permitting the movement of data in this way: *drag-and-drop transfer of messages.*

DRAM /ˈdē ˌræm/ ▸ **abbr.** dynamic random-access memory.

drill /dril/ ▸ **v.** [intrans.] (**drill down**) access data that is in a lower level of a hierarchically structured database: *just as the department can view a single screen showing the IT infrastructure, and drill down to any faulty component, so a single screen shows the entire electricity distribution infrastructure, and can drill down to the source of any fault.*

drive /drīv/ ▸ **n.** short for **DISK DRIVE**.

drive bay /ˈdrīv ˌbā/ ▸ **n.** a space in a computer case where a disk drive can be accommodated.

driv•er /ˈdrīvər/ ▸ **n.** a program that controls the operation of a device such as a printer or scanner.

DERIVATIVES **driv•er•less adj.**

droid /droid/ ▸ **n.** a program that automatically collects information from remote systems.

ORIGIN 1970s: shortening of *android*.

drop-down /ˈdräp ˌdown/ ▸ **adj.** [attrib.] (of a menu) appearing below a menu title when it is selected, and remaining until used or deselected. Compare with **PULL-DOWN**.

drop•out /ˈdräpˌowt/ ▸ **n.** a momentary loss of recorded audio signal or an error in reading data on a magnetic tape or disk, usually due to a flaw in the coating.

DSL ▸ **abbr.** digital subscriber line, a technology for delivering continuous high-speed Internet access by digital transmission over a telephone line while maintaining normal telephone service on the same line.

DSS ▸ **abbr.** digital signature standard.

DTD ▸ **abbr.** document type definition, a template that sets out the tag structure of an electronically created document using a markup language such as SGML, XML, or HTML.

dub-dub-dub /ˈdəb ˈdəb ˈdəb/ ▸ **n.** informal short form used instead of pronouncing the three letters in the abbreviation WWW (World Wide Web).

dumb /dəm/ ▸ **adj.** (of a computer terminal) able only to transmit data to or receive data from a computer; having no independent processing capability. Often contrasted with **INTELLIGENT**.

ORIGIN Old English, of Germanic origin; related to Old Norse *dumbr* and Gothic *dumbs* 'mute,' also to Dutch *dom* 'stupid' and German *dumm* 'stupid.'

dump /dəmp/ ▸ **n.** a copying of stored data to a different location, performed typically as a protection against loss.

■ a printout or list of the contents of a computer's memory, occurring typically after a system failure.

▸ **v.** [trans.] copy (stored data) to a different location, esp. so as to protect against loss: *a hard disk that you can dump your pictures to.*

■ print out or list the contents of a computer's memory, esp. after a system failure.

ORIGIN Middle English: perhaps from Old Norse; related to Danish *dumpe* and Norwegian *dumpa* 'fall suddenly' (the original sense in English); in later use partly imitative.

Dun•geons and Drag•ons /ˈdənjənz ən ˈdragənz/ ▸ **trademark** a fantasy role-playing electronic game set in an imaginary world based loosely on medieval myth.

du•plex /ˈd(y)o͞opleks/ ▸ **adj. 1** (also **full du•plex** /ˈfo͝ol ˈd(y)o͞opleks/) (of a communications device, computer circuit, etc.) allowing the transmission of two signals simultaneously in opposite directions: *full duplex communication.* Also see **HALF-DUPLEX**.

2 (of a printer or its software) capable of printing on both sides of the paper.

ORIGIN mid 16th cent. (as an adjective): from Latin *duplex, duplic-*, from *duo* 'two' + *plicare* 'to fold.'

DVD ▸ **n.** a small plastic high-density disc that stores large amounts of data, esp. high-resolution audio-visual material, and from which the information can be read using reflected laser light.

ORIGIN abbreviation of *digital versatile disc* or *digital videodisc*.

USAGE: The many different DVD formats can be very confusing. A plain-vanilla DVD (sometimes called a DVD-ROM, on the model of CD-ROM) can be played in DVD players and by computers that have a DVD drive, but cannot be written to by the user. A DVD-R can also be played in DVD players and by computer DVD drives, and can be written to, but only once, by a home DVD burner. DVD-Rs come in two sizes; 4.7 GB and 9.4 GB. DVD-RAM discs can be written to many times, but can only be played by a computer's DVD-RAM drive. They cannot be played in other DVD players or drives. DVD-RW discs can be written to as many as a thousand times, but may not play in some older or lower-level DVD players or drives. DVD+R and DVD+RW discs are much like DVD-R discs, but only come in the 4.7 GB size. DVD+RW discs are also unlikely to play back on older or lower-level DVD players.

DVD-R ▸ **n.** a blank DVD on which data, including music and movies, can be permanently recorded and read using the DVD-R format. See usage note at **DVD.**

■ a format for recordable DVDs used by some companies.

ORIGIN an abbreviation of *DVD recordable.*

DVD+R ▸ **n.** a blank DVD on which data, including music and movies, can be permanently recorded and read using the DVD+R format.

■ a format for recordable DVDs used by some companies. See usage note at **DVD.**

ORIGIN an abbreviation of *DVD plus recordable.*

DVD-RAM ▸ **n.** a blank DVD, enclosed in a cartridge, on which data, including music and movies, can be permanently recorded and read using the DVD-RAM format. DVD-RAM discs can be recorded over many times, but will only play back in a DVD-RAM drive.

■ a format for recordable DVDs used by some companies. See usage note at **DVD.**

ORIGIN an abbreviation of *DVD random access memory.*

DVD-ROM ▸ **n.** a DVD used as a read-only optical memory device for a computer system. See usage note at **DVD.**

ORIGIN 1990s: acronym from *DVD read-only memory.*

DVD-RW ▸ **n.** a blank DVD that can be recorded, erased, and re-recorded with data many times and read by systems using the DVD-RW format.

■ a format for rewritable DVDs used by some companies. ■ a disc drive that can read and record DVDs. See usage note at **DVD.**

ORIGIN an abbreviation of *DVD rewritable.*

DVD+RW ▸ **n.** a blank DVD that can be recorded, erased, and re-recorded with data many times and read by systems using the DVD+RW format.

■ a format for rewritable DVDs used by some companies. ■ a disc drive that can read and record DVDs. See usage note at **DVD.**

ORIGIN an abbreviation of *DVD plus rewritable*.

dy•nam•ic /dī'næmik/ ▸ **adj.** (of a memory device) needing to be refreshed by the periodic application of a voltage.

DERIVATIVES **dy•nam•i•cal•ly adv.**

ORIGIN early 19th cent. (as a term in physics): from French *dynamique*, from Greek *dunamikos*, from *dunamis* 'power.'

dy•nam•ic link li•brar•y /dī'næmik 'liNGk ˌlībrerē; -brərē/ (abbreviation: **DLL**) ▸ **n.** see **DLL.**

BASIC TROUBLESHOOTING

Is your computer a little problematic? Slowing down? And you're not ready to call in expert help? Try these tricks. As always, back up important data first.

General tips for Mac and Windows systems

1. A surprising number of computer problems—inexplicable system crashes, random program shutdowns, "not enough memory" errors—are caused by faulty RAM chips. These errors are normally hard to diagnose by yourself, but on Windows, the free program Memtest86 (http://www.mem test86.com) makes it easy. On Macintosh OS 9, use RAM Check 2.1, which can be downloaded at http://ftp.traffictrak.com/RAMCheck21.sit or http://download.digidesign.com/support/digi/mac/utilities/RMCheck210.sea.hqx(.) You can also try Techtool Pro for OS 9 or OS X.

2. Four big demons of computers are dust, heat, moisture, and static. You can take some preventive measures, like covering your CPU and monitor when they're not in use and protecting your computer by drinking your coffee away from the computer, but demons will strike no matter what you do, and they manifest their presence in odd ways. The most common is disk corruption, which shows up as disappearing or corrupt files and folders. Regular disk maintenance is important to keep these demons at bay.

The most well-known tool for repairing disk drives is Norton Utilities (http://www.symantec.com), which is included in Norton System Works. You can use it to diagnose and fix already corrupt files, but it is most useful for *preventive* maintenance. Using these utilities to optimize or defragment your hard drive on a regular basis makes individual files as contiguous as possible, which can increase the speed at which files and programs are opened.

However, Norton Utilities can sometimes cause problems of its own, particularly on older computers, if it is installed to run in the background. If you open System Works and click on Options, Norton Utilities, and then the Startup tab, you can uncheck the boxes that would instruct the program to start whenever you boot up. Then it's up to you to decide how often you need to activate it, perhaps somewhere between once a week and once a month.

Other drive utilities to consider for Windows are Avantrix Utilities Suite (http://www.avantrix.com), WinRescue (http://www.superwin.com/index.htm), and TuneUp Utilities (http://www.tune-up.com). These programs will also clean the registry and remove cache or temporary files.

For Macintosh, DiskWarrior (http://www.alsoft.com/DiskWarrior) is the best directory-repair program for OS 9 and OS X. You can also try Drive10 and the aforementioned TechTool Pro (http://www.micromat.com).

3. Buy more RAM; RAM is cheap and fast. The more you have, the less time your computer spends reading and writing information to the hard drive instead of using RAM. RAM comes in DIMMs (dual in-line memory modules) and SIMMs (single in-line memory modules), small circuit boards that hold the RAM chips. Before you buy either, find out what kind of modules are already installed in your computer (you may have to replace, not just add), how much RAM your motherboard can hold if you "max it out," and which manufacturer makes boards compatible with your computer. Don't panic about how to get system infor-

mation. A shareware program called Sandra that analyzes your computer system is available at http://www.sisoftware.com(.) Once you have the information, reliable name-brand manufacturers like Crucial (http://www.crucial.com) and Kingston (http://www.kingston.com) allow you to specify the brand and model number of your computer so you can find exactly what you need. Of course you can order RAM directly from the manufacturer of your computer, but chances are you'll pay more.

4. Keep the operating system current by downloading the latest fixes and updates. A high-speed Internet connection will keep downloads from taking forever, but dial-up will still work and updating is important. In Windows, run Windows Update (usually in the Start menu) and make sure to get all updates marked as "critical." On Macintosh OS 9 and OS X, run Software Update, found in the control panels or system preferences.

Tips Specific to Microsoft Windows

1. *Get antivirus software, activate it, and keep it up to date.* Running a computer without virus protection is like walking around without your skin: close contact with others is bound to expose you to infection sooner or later.

Windows is particularly vulnerable to virus attacks, especially if Outlook or Outlook Express is your e-mail client. Although viruses can contaminate your computer through tangible objects (e.g., someone else's floppies or CDs), the primary avenues for infection are downloaded files (are you still using peer-to-peer file sharing?) and e-mail. You can't be warned often enough: do not open e-mail attachments with the extensions .bat, .exe, .scr, .zip, .pif, or .com unless you know that someone reliable has sent them.

Many computers are sold with only a trial version of antivirus software, which soon expires. As a result, you may believe your computer is protected when it no longer is. Good packages, well worth the money, are provided by Symantec (http://symantec.com), which makes Norton Antivirus, McAfee (http://www.mcafee.com), F-Secure (http://www.f-secure.com), and Sophos (http://www.sophos.com). A free antivirus program that works well is offered by Grisoft (http://www.grisoft.com). If you have an "always on" connection to the Internet, set the program to reach out and touch you by letting it update your virus definitions automatically; otherwise, run the updater manually every day. At least once a week, run a full virus scan. Many viruses now send e-mails purporting to be from someone you know; if you are not expecting an attachment or if it comes with a cryptic message, delete it.

2. *Capture spies.* Strange window (that's "window," not "Windows") behavior, mysteriously appearing ads, crashing programs, strange tool bars in your web browser, unusual error messages, and slow performance are some of the symptoms that indicate that you have spyware on your system. Spyware installs itself automatically when you visit certain web pages or piggybacks on more legitimate programs.

Besides using your computer's memory, CPU cycles, and hard drive space, spyware may also be tracking which web pages you visit, replacing paid-for ads on web sites with its own, or even using your computer to operate as a spam relay. These are not trivial matters.

To remove spyware—there's no good reason for it to be there—run SpyBotSD (SD = Search and Destroy!), available at http://www.safer-networking.org(.) This free program will detect and remove dozens of spyware programs, leaving you with a more efficient and secure system. But proceed with caution. SpyBot can be excessively eager; the program that runs *The New Oxford American Dictionary* CD, for example, is sometimes wrongly accused. Don't remove it!

3. *Defeat sluggish performance by removing unnecessary programs from your startup file.* Do you have a lot of icons in the Windows system tray in the lower right hand corner of your screen? Some of them may represent aggressive programs that have

been thrust into your startup file without your permission, even though you use them only occasionally. Each program there lurks in the background every time you boot up and takes up memory and processor cycles, whether or not it is in use. Right-click on each icon to learn what the program does. Any that you don't need at startup can safely be deactivated; it will still be available for use. Good programs for allowing you to control which software should load automatically when you boot up is Startup Copilot ($20), found at http://www.windows-startup-cop.com(,) or the new Startup Cop Pro, which you can get through http://www.pcmag.com/article2/0,4149, 1406616,00. asp(.)

4. *Use Windows Safe Mode as a diagnostic tool.* If you're getting an error when you turn your computer on, Safe Mode, which loads only the components that run the operating system, can allow you to disable programs that ordinarily run at startup. You can disable them one at a time, through the Control Panel and then test to see which one is causing the problem. Instructions for invoking and using Safe Mode in the various versions of the Windows operating system can be found at http://service1.symantec.com/SUPPORT/tsgeninfo.nsf/docid/2001052409420406?OpenDocument&src=sec_doc_nam(.) Come to think of it, you might want to check this site out *before* there's a problem, so you're prepared if your computer boots into Safe Mode on its own.

Apple Macintosh

Although Macs generally have fewer problems of this nature than Windows machines, they do have their issues.

1. In OS 9, holding down the Option and Apple keys during bootup will rebuild the Desktop. This bit of minor maintenance is worth doing once a month or so; it keeps files and icons associated with their appropriate applications.

2. In recent versions of OS X, running the Disk Utility, choosing the First Aid tab, and clicking on Repair Disk Privileges

resolves a remarkable number of issues, as when you've received an "access denied" message. Because the underlying Unix layer is so sensitive to changes in rights and privileges, you should do this after every system update.

3. If you're getting an error on startup that prevents you from working, you can, in either OS 9 or OS X, hold down the Shift key on bootup to disable all nonessential extension and startup items. While you can't work in this mode, it allows you to get to your files until you can determine which program or extension is causing the problem.

4. In OS 9.1 and above, you should turn on virtual memory. This allows for more efficient use of RAM.

E

e /ē/ ▸**n.** (pl. **e's**) an e-mail system, message, or messages: *On your next business trip, you might want to go surfing, or you may want to do an "e." An e-mail that is.*

▸**v.** (**e'd, e'•ing**) [trans.] **1** send an e-mail to (someone): *Anime Vids FOR SALE (E-me to make offer).*

2 send (a message) by e-mail.

e- ▸ **prefix** denoting anything in an electronic state, esp. the use of electronic data transfer in cyberspace for information exchange and financial transactions, esp. through the Internet: *e-business* | *e-commerce* | *e-world* | *e-zine.*

ORIGIN from **ELECTRONIC**, on the pattern of *e-mail.*

EBCDIC /ˈebsēˌdik/ ▸ **abbr.** Extended Binary Coded Decimal Interchange Code, a standard eight-bit character code used in computing and data transmission.

e-book /ˈē ˌbo͝ok/ ▸**n.** an electronic version of a printed book that can be read on a personal computer or hand-held device designed specifically for this purpose: *the arrival of the e-book could change the way many books are read, written and published—a shakeup under way thanks to the Internet.* ■ a dedicated device for reading electronic versions of printed books.

ech•o /ˈekō/ ▸ **v.** (**ech•oes, ech•oed**) [trans.] send a copy of (an input signal or character) back to its source or to a screen for display: *for security reasons, the password will not be echoed to the screen.*

ORIGIN Middle English: from Old French or Latin, from Greek *ēkhō*, related to *ēkhē* 'a sound.'

e•con•tent /ˈēˌkäntent/ (also **e-con•tent, e•Con•tent**) ▸**n.** text and images designed for display on web pages: [often as modifier] *Visit* *our eContent call pages for more details of the multilingual and multicultural content action line.*

EDI ▸ **abbr.** electronic data interchange (a standard for exchanging information between computer systems).

ed•it /ˈedit/ ▸ **v.** (**ed•it•ed, ed•it•ing**) [trans.] (often **be edited**) correct, condense, or otherwise modify (data or software).

■ (**edit something out**) remove unnecessary or inappropriate words, sounds, scenes, etc., from a text, sound file, photo, graphic, movie or animation.

▸ **n.** a change or correction made as a result of editing.

ORIGIN late 18th cent. (as a verb): partly a back-formation from **EDITOR**, reinforced by French *éditer* 'to edit' (from *édition* 'edition').

ed•it•a•ble /ˈedit̠əbəl/ ▸ **adj.** (of text, graphics, software, etc.) in a format that can be edited by the user: *editable graphics | the software converts a scanned document to editable text.*

ed•i•tor /ˈedit̠ər/ ▸ **n.** a computer program enabling the user to alter or rearrange online text.

■ a person who is in charge of and determines the final content of a text.

ORIGIN mid 17th cent.: from Latin, 'producer (of games), publisher,' from *edit-* 'produced, put forth,' from the verb *edere.*

EDP ▸ **abbr.** electronic data processing.

E-fit /ˈē ˌfit/ ▸ **n.** an electronic picture of a person's face made from composite photographs of facial features, created by a computer program: *the suspects were identified to police from computer-enhanced photofits, or E-fits, of the two men.*

ORIGIN 1980s: from *e-* 'electronic' and *fit*, on the pattern of *Photofit* (trademark method of creating a composite picture of a crime suspect's face).

EGD ▸ **n.** a technology or system that integrates a computer display with a pair of eyeglasses, using a lens or mirror to reflect images into the eyes: *some EGDs are designed to clip right on to your eyeglasses.*

ORIGIN abbreviation of 'eyeglass display.'

e•go•surf /ˈēgōˌsərf/ ▸ **v.** [intrans.] informal search the Internet for instances of one's own name or links to one's own Web site: *I ego-surfed a bit to see what the DSS would find if they searched the Web to find out about me.*

DERIVATIVES **e•go•surf•ing n.**

EIDE ▸ abbr. enhanced integrated drive electronics. See **IDE**.

e•lec•tron•ic /iˌlek'tränik/ ▸ **adj. 1** relating to or carried out using a computer or other electronic device, esp. over a network: *electronic banking.*

2 (of a device) having, or operating with the aid of, many small components, esp. microchips and transistors, that control and direct an electric current: *an electronic device.*

DERIVATIVES **e•lec•tron•i•cal•ly adv.**

ORIGIN early 20th cent.: from *electron* + *-ic.*

e•lec•tron•ic mail /iˌlek'tränik 'māl/ ▸ **n.** another term for **E-MAIL**.

e•lec•tron•ic or•gan•iz•er /iˌlek'tränik 'ôrgəˌnīzər/ ▸ **n.** a pocket-sized computer used for storing and retrieving information such as addresses and appointments: *that's right, I don't believe in Filofaxes or electronic organizers or datebooks or any of those other sissy devices for making a list.*

e-mail /'ē ˌmāl/ (also **e•mail**) ▸ **n.** messages distributed by electronic means from one computer user to one or more recipients via the Internet or a local network: *reading e-mail has become the first task of the morning* | [as adjective] *e-mail messages.*

■ the system of sending messages by such electronic means: *a contract communicated by e-mail.*

▸ **v.** [trans.] send an e-mail to (someone): *you can e-mail me at my normal address.*

■ send (a message) by e-mail: *employees can e-mail the results back.*

DERIVATIVES **e-mail•er n.**

ORIGIN late 20th cent.: abbreviation of **ELECTRONIC MAIL**.

e-mail ad•dress /'ē ˌmāl əˌdres/ ▸ **n.** a string of characters comprising an e-mail name, the symbol @, and the domain name of the e-mail service provider, which directs e-mail to the correct user.

em•bed /em'bed/ (also **im•bed** /im'bed/) ▸ **v.** (**em•bed•ded, em•bed•ding**) [trans.] (often **be embedded**) **1** incorporate (a text or code) within the body of a file or document: [as adjective] *an embedded link.*

■ incorporate (a program) into a device: *the company plans to embed browsing software in the new memory cards.*

2 [often as adjective] (**embedded**) design and build (a microprocessor) as an integral part of a system or device.

DERIVATIVES **em•bed•ment n.**

e•mo•ti•con /i'mōṯi,kän/ ▸ **n.** a representation of a facial expression such as :-) (representing a smile), formed by various combinations of keyboard characters and used in electronic communications to convey the writer's feelings or intended tone.

ORIGIN 1990s: blend of *emotion* and **ICON.**

EMS ▸ **abbr.** expanded memory system, a system for increasing the amount of memory available to a personal computer, now largely superseded by XMS.

em•u•late /'emyə,lāt/ ▸ **v.** [trans.] reproduce the function or action of (a different computer or software system): *users are able to emulate one operating system from within another.*

DERIVATIVES **em•u•la•tion n.**

ORIGIN late 16th cent.: from Latin *aemulat-* 'rivaled, equaled,' from the verb *aemulari*, from *aemulus* 'rival.'

em•u•la•tor /'emyə,lātər/ ▸ **n.** (also **em•u•la•tor pro•gram**) a program enabling a computer to execute programs written for a different computer.

DERIVATIVES **em•u•la•tion** /,emyə'lāsHən/ **n.**

ORIGIN late 16th cent.: from Latin *aemulat-* 'rivaled, equaled,' from the verb *aemulari*, from *aemulus* 'rival.'

en•a•ble /en'ābəl/ ▸ **v.** [trans.] make (a device or system) operational; activate: *enable file sharing.*

DERIVATIVES **en•a•ble•ment n.**

en•a•bled /en'ābəld/ ▸ **adj.** [in combination] adapted for use with the specified application or system: *Java-enabled push technology which aims to make the Internet a simple, yet media-rich, plug 'n' play environment.*

en•cap•su•late /en'kæps(y)ə,lāt/ ▸ **v.** [trans.] enclose (a message or signal) in a set of codes that allow use by or transfer through different computer systems or networks.

■ provide an interface for (a piece of software or hardware) to allow or simplify access for the user.

DERIVATIVES **en•cap•su•la•tion** /enˌkæps(y)əˈlāsHən/ **n.**

en•code /enˈkōd/ ▸ **v.** [trans.] convert (information or an instruction) into a digital form: *the standard used to encode and play digital sound.* Compare with **ENCRYPT**.

DERIVATIVES **en•cod•a•ble** adj.; **en•cod•er** n.; **en•code•ment** n.

en•crypt /enˈkript/ ▸ **v.** [trans.] convert (information or data) into a cipher or code, esp. to prevent unauthorized access: *a new feature is the ability to encrypt files so only you can use them.*

■ (**encrypt something in**) conceal information or data in something by this means. Compare with **ENCODE**.

DERIVATIVES **en•cryp•tion** /enˈkripsHən/ **n.**

ORIGIN 1950s (originally U.S.): from *en-* 'in' + Greek *kruptos* 'hidden.'

end•i•an /ˈendēən/ ▸ **adj.** denoting or relating to a system of ordering data in a computer's memory whereby the most significant (**big endian**) or least significant (**little endian**) byte is put first: *Microsoft recently confirmed that Virtual PC does not work on G5 processors, because processor support for a little endian mode does not exist on the G5.*

ORIGIN 1980s: a reference to Swift's *Gulliver's Travels*, in which the Lilliputians were divided into two camps, those who ate their eggs by opening the 'big' end and those who ate them by opening the 'little' end.

end us•er /ˈend ˌyo͞ozər/ (also **end-us•er**) ▸ **n.** the person who actually uses a particular product.

en•queue /enˈkyo͞o/ ▸ **v.** [trans.] (**en•queued**, **en•queu•ing** or **en•queue•ing**) add (an item) to a queue: *your message has been enqueued and undeliverable for 1 day(s) to the following recipients.*

en•ter /ˈentər/ ▸ **v.** [trans.] type (information) in a computer so as to record it: *children can **enter** the data **into** the system.*

▸ **n.** (also **en•ter key**) on some computer keyboards, a key that is used to perform various functions, such as executing a command or selecting options on a menu.

ORIGIN Middle English: from Old French *entrer*, from Latin *intrare*, from *intra* 'within.'

en•vi•ron•ment /en'vīr(ə)nmənt/ ▸ **n.** [with adjective] the overall structure within which a user, computer, or program operates: *a desktop development environment.*

e•rase /i'rās/ ▸ **v.** [trans.] remove (data) from (a disk); delete (data) from a computer's memory.

DERIVATIVES **e•ras•a•ble adj.**; **e•ra•sure** /i'rāSHər/ **n.**

ORIGIN late 16th cent. (originally as a heraldic term meaning 'represent the head or limb of an animal with a jagged edge'): from Latin *eras-* 'scraped away,' from the verb *eradere*, from *e-* (variant of *ex-*) 'out' + *radere* 'scrape.'

er•ror cor•rec•tion /'erər kə,rekSHən/ ▸ **n.** the automatic correction of errors that arise from the incorrect transmission of digital data.

er•ror mes•sage /'erər ,mesij/ ▸ **n.** a message displayed on a screen or printout, indicating that an incorrect instruction has been given, or that there is an error resulting from faulty software or hardware.

es•cape /i'skāp/ ▸ **n.** (also **es•cape key** /i'skāp ,kē/) a key on a computer keyboard that either interrupts the current operation or converts subsequent characters to a control sequence.

ORIGIN Middle English: from Old French *eschaper*, based on medieval Latin *ex-* 'out' + *cappa* 'cloak.'

E•ther•net /'ēTHər,net/ ▸ **trademark** a system for connecting a number of computers to form a local area network, with protocols to control the passing of information and to avoid simultaneous transmission by two or more systems: *high-speed Ethernet* | [as adj.] *a built-in Ethernet port.*

■ a network using this.

ORIGIN 1970s: blend of *ether* and **NETWORK**.

EULA /'ē'yo͞o'el'ā; 'yo͞olə/ ▸ **abbr.** end user license agreement.

e•vent cre•a•tion /i'vent krē,āSHən/ ▸ **n.** (in computer programming) the activity of or facility for creating an event that will unfold in real-time when conditions for it have been met: *I have tried using the createEventProc method for event creation but my latest effort has been using the insertLine method of the codeModule of interest.*

ex•e•cut•a•ble /'eksi,kyo͞otəbəl/ ▸ **adj.** (of a file or program) able to be run by a computer: *executable code in e-mail attachments.*

▸ **n.** an executable file or program: *an e-mail with an attachment that contains an executable.*

ex•e•cute /'eksi,kyo͞ot/ ▸ **v.** [trans.] carry out an instruction or run a program.

ORIGIN late Middle English: from Old French *executer*, from medieval Latin *executare*, from Latin *exsequi* 'follow up, carry out, punish,' from *ex-* 'out' + *sequi* 'follow.'

ex•e•cu•tion /,eksi'kyo͞osHən/ ▸ **n.** the performance of an instruction or a program.

EXE file /'eksē ,fīl; 'ē'eks'ē/ ▸ **n.** an executable program file having the extension .exe.

ex•it /'egzit; 'eksit/ ▸ **v.** (**ex•it•ed**, **ex•it•ing**) terminate a process or program, usually returning to an earlier or more general level of interaction: [trans.] *save the document and exit the program.*

ex•pan•sion card /ik'spænsHən ,kärd/ (also **ex•pan•sion board**) ▸ **n.** a circuit board that can be inserted in a computer to give extra facilities or memory.

ex•pan•sion slot /ik'spænsHən ,slät/ ▸ **n.** a place in a computer where an expansion card can be inserted.

ex•pert sys•tem /'ekspərt ,sistəm/ ▸ **n.** a piece of software programmed using artificial intelligence techniques. Such systems use databases of expert knowledge to offer advice or make decisions in such areas as medical diagnosis and trading on the stock exchange.

ex•port /ik'spôrt; 'ekspôrt/ ▸ **v.** [trans.] transfer (data) in a format that can be used by other programs: *you can export the address book to use in another application.*

DERIVATIVES **ex•port•a•bil•i•ty n.**; **ex•port•a•ble adj.**

ex•ten•sion /ik'stensHən/ ▸ **n.** an optional suffix to a file name, typically consisting of a period followed by several characters, indicating the file's content or function. Extensions are up to three characters long and are optional, but help to identify whether the file is a basic program, .BAS; a program stored on disk in a ready-to-use form, .COM or .EXE; a spreadsheet calculation file, .CAL; or a word-processing document, .DOC.

ex•ter•nal /ik'stərnl/ ▸ **adj.** (of hardware) not contained in the main computer; peripheral: *an external floppy drive.*

ORIGIN late Middle English: from medieval Latin, from Latin *exter* 'outer.'

ex•tra•net /ˈekstrəˌnet/ ▸ **n.** an intranet that can be partially accessed by authorized outside users, enabling businesses to exchange information over the Internet in a secure way: *this takes information from the SQL server and publishes it to an intranet for Allianz Indonesia staff, and to a secure extranet for brokers and agents.*

ORIGIN 1990s: from *extra-* 'outside' + *net*, by analogy with *intranet.*

eye track•ing /ˈī ˌtrakiNG/ (also **eye-track•ing**) ▸ **n.** a technology that monitors eye movements as a means of detecting abnormalities or of studying how people interact with text or online documents: *a privately-held company that uses eye tracking to evaluate visual products.*

e-zine /ˈē ˌzēn/ ▸ **n.** a magazine only published in electronic form on a computer network.

F

fab /fæb/ ▸ **n.** Electronics a microchip fabrication plant: *with about 900 fabs now operating in the world and an attrition rate of 20 years, the industry ought to be replacing about 40 to 50 plants a year.*

■ a particular fabrication process in such a plant.

ORIGIN late 20th cent.: abbreviation of *fabrication.*

fab•less /ˈfæblis/ ▸ **adj.** denoting or relating to a company that designs microchips but contracts out their production rather than owning its own factory: *the newcomers' strategy was fabless production. . . . Let the Japanese make the stuff; we'll design it and reap most of the profit.*

ORIGIN 1980s: from *fab* 'a microchip fabrication plant' + *-less.*

face•print /ˈfāsˌprint/ ▸ **n.** a digital scan or photograph of a human face, used for identifying individuals from the unique characteristics of facial structure: *hidden cameras and faceprints are used to single out individuals in a crowd.*

ORIGIN on the pattern of *fingerprint.*

face•print•ing /ˈfāsˌprintiNG/ ▸ **n.** the process of creating a digital faceprint and using software to compare it with a database of photographs, especially to identify known criminals: *the "biometrics" industry, of which faceprinting is only a part, is now trying to find a few heart-warming "Good News Biometric Stories" to counteract opposition.*

ORIGIN on the pattern of *fingerprinting.*

fail•o•ver /ˈfālˌōvər/ ▸ **n.** a method of protecting computer systems from failure, in which standby equipment automatically takes over when the main system fails: *automatic failovers can be handled in* *one of two ways—either via mirroring software or through shared-disk clustering.*

false col•or /ˈfôls ˈkələr/ ▸ **n.** color added during the processing of a photographic or computer image to aid interpretation of the subject: [as adj] *a false-color image created from satellite data.*

FAQ /fæk/ ▸ **n.** a text file containing a list of questions and answers relating to a particular subject, esp. one giving basic information for users of an Internet newsgroup.

ORIGIN 1990s: acronym from *frequently asked questions.*

field /fēld/ ▸ **n.** a part of a record, representing an item of data.

FIFO /ˈfīˌfō/ ▸ **abbr.** first in, first out (with reference to methods of data storage). Compare with **LIFO**.

fifth-gen•er•a•tion com•put•er /ˈfifTH ˌjenəˈrāSHən kəmˈpyo͞otər/ ▸ **n.** a proposed new class of computer employing artificial intelligence.

file /fīl/ ▸ **n.** a collection of data, programs, etc., stored in a computer's memory or on a storage device under a single identifying name: *do you want to save this file?* | [as adjective] *a file name.*

DERIVATIVES **fil•er n.**

ORIGIN late Middle English (as a verb meaning 'string documents on a thread or wire to keep them in order'): from French *filer* 'to string,' *fil* 'a thread,' both from Latin *filum* 'a thread.'

file ex•ten•sion /ˈfīl ikˌstenSHən/ ▸ **n.** see **EXTENSION.**

file•name /ˈfīl nām/ (also **file name**) ▸ **n.** a name given to a computer file, often including an extension that identifies the particular type of file: *the virus creates copies of itself with random filenames.* See **EXTENSION.**

USAGE: What's in a filename? You can name your child anything you like, but if you want your computer to handle a file correctly, you should follow some rules when naming your files. It's best to use only lower-case letters in your filenames, especially if you are going to be putting them up on your Web site. (Macs and Windows systems treat "MyDoc.txt" and "mydoc.txt" the same way, but on Unix systems, they're two different files!) Macs and PCs allow spaces in filenames; Unix machines do not. Make sure your filename is descriptive—"letter.doc" won't mean as much to you in a month as "amy_letter_may.doc will. Avoid using non-alphabetic characters, especially slashes and colons. (Using the underscore character is fine.) Especially on a PC, it's always a good idea to use the right extensions. Most programs will save files with the right extension automatically; all you have to do is remember to override the function.

file serv•er /ˈfīl ˌsərvər/ ▸ **n.** a device that controls access to separately stored files, as part of a multiuser system.

file-shar•ing /ˈfīl ˌSHe(ə)riNG/ ▸ **n.** the transmission of files from one computer to another over a network or the Internet: [often as modifier] *a judge has questioned whether distributors of file-sharing software programs are responsible for illegal trading of copyrighted material.*

fill /fil/ ▸ **n.** the shading in of a region of a computer graphic.

ORIGIN Old English *fyllan* (verb), *fyllu* (noun) of Germanic origin; related to Dutch *vullen* and German *füllen* (verbs), *Fülle* (noun).

fil•ter /ˈfiltər/ ▸ **n.** a piece of software that processes text, for example to remove unwanted spaces or to format it for use in another application.

■ a set of instructions or piece of software that processes and sorts incoming data according to defined criteria: *an e-mail filter.*

▸**v.** [trans.] process (data) to remove unwanted material: *the program lets you filter incoming mail.*

fire•wall /ˈfī(ə)rˌwôl/ ▸ **n.** a system of software features that prevent unauthorized access to a computer via a network: *all requests from inside the firewall first go to the proxy server, which then makes the request to the external Internet. The proxy server retrieves the data and returns it to the computer that made the original request inside the firewall.*

▸**v.** protect with a firewall: *the ability to firewall broadcast frames is the switch's most important feature.*

FireWire /ˈfī(ə)rˌwī(ə)r/ ▸ **trademark** a technology that allows high-speed communication and data exchange between a computer and a peripheral or between two computers: *transfer of digital content to DVD over FireWire* | [as adj.] *a FireWire port.* Also see **IEEE 1394**.

firm•ware /ˈfərmˌwe(ə)r/ ▸ **n.** permanent software programmed into a read-only memory.

fix /fiks/ ▸ **n.** informal a solution to a problem, esp. one that is hastily devised or makeshift: *the company released a quick fix for the bug, but a better solution will have to wait for the next release.*

fixed-point /ˈfikst ˈpoint/ ▸ **adj.** denoting a mode of representing a number by a single sequence of digits whose values depend on their location relative to a predetermined base point. In a fixed-point notation, only a set number of digits can follow the decimal point: *these computers perform arithmetic in fixed-point binary format.* Often contrasted with **FLOATING-POINT**.

flag /flæg/ ▸ **n.** a variable used to indicate a particular property of the data in a record.

▸ **v.** (**flagged**, **flag•ging**) [trans.] (often **be flagged**) mark (an item) for attention or treatment in a specified way: *any records that have changed will be flagged in the database.*

ORIGIN mid 16th cent.: perhaps from obsolete *flag* 'drooping,' of unknown ultimate origin.

flame /flām/ ▸ **n.** a vitriolic or abusive message posted to a message board or sent via e-mail, typically in quick response to another message: *flames about inexperienced users posting stupid messages.*

▸ **v.** [trans.] send (someone) abusive or vitriolic e-mail messages, typically in a quick exchange.

DERIVATIVES **flam•er n.**

flame war /ˈflām ˌwôr/ ▸ n. an extended exchange involving two or more people of abusive or vitriolic electronic messages sent to a message board or e-mail group.

flam•ing /ˈflāmiNG/ ▸ **adj.** [attrib.] (of electronic messages) abusive or vitriolic.

DERIVATIVES **flam•ing•ly adv.**

flash drive /ˈflæSH ˌdrīv/ ▸ **n.** a data storage device containing flash memory that has no moving parts and does not need batteries or a power supply. Also see **USB FLASH DRIVE**.

flash mem•o•ry /ˈflæSH ˌmem(ə)rē/ ▸ **n.** solid-state memory that can be erased and reprogramed and retains data in the absence of a power supply: *the diagnostics are kept in flash memory* | [as adjective] *a flash memory card.*

flat file /ˈflæt ˈfīl/ ▸ **n.** a file having no internal hierarchy.

■ [as adjective] denoting a system using such files: *a flat-file database.* ■ a file from which all special characters or word processing instructions have been removed: *send us a flat file and we'll reformat it in our system.*

flat-pan•el mon•i•tor /'flæt ˌpænl 'mänit̲ər/ ▸ **n.** (also **flat-pan•el dis•play** /'flæt ˌpænl di'splā/, **flat panel**) a liquid crystal computer monitor that is narrow in depth.

flat-screen mon•i•tor /'flæt ˌskrēn 'mänit̲ər/ ▸ **n.** (also **flat-screen dis•play** /'flæt ˌskrēn di'splā/) a CRT computer monitor with a flat-faced tube.

flip /flip/ ▸ **v.** (**flipped, flip•ping**) [trans.] access the nonpublic parts of (a Web site): *if you want to learn who the main IT contact at a company is, just flip their Web site.*

flip-chip /'flipˌCHip/ ▸ **n.** a computer chip that is installed on a circuit board face down, with connections formed by solder bumps rather than wires: *flip chips offer the possibility of low cost electronic assembly for modern electronic products because interconnection on the chip can be made simultaneously in a single step.*

ORIGIN 1990s: from the fact the the chip is rotated 180 degrees from the traditional mode of attachment.

float•ing-point /'flōtiNG ˌpoint/ ▸ **adj.** denoting a mode of representing numbers as two sequences of bits, one representing the digits in the number and the other an exponent that determines the position of the base point. In a floating-point notation there can be any number of digits following the decimal point: *speeds of more than one million* ***floating-point operations*** *per second.* Often contrasted with FIXED-POINT.

-flop /fläp/ ▸ **comb. form** floating-point operations per second (used as a measure of computing power): *a gigaflop computer.*

ORIGIN acronym; originally spelled *-flops* (*s* = second) but shortened to avoid misinterpretation as plural.

flop•py /'fläpē/ ▸ **n.** (pl. **flop•pies**) short for FLOPPY DISK.

flop•py disk /'fläpē 'disk/ ▸ **n.** a flexible removable magnetic disk, typically encased in hard plastic, used for storing data. Also called DISKETTE. Compare with HARD DISK.

flop•ti•cal /'fläptikəl/ ▸ **trademark** denoting or relating to a type of floppy-disk drive using a laser to position the read-write head.

▸ **n.** a floppy-disk drive of this type.

ORIGIN 1980s: blend of **FLOPPY** and **OPTICAL**.

flow chart /ˈflō ˌCHärt/ (also **flow•chart** or **flow di•a•gram** /ˈdīə ˌgræm/) ▸ **n.** a graphical representation of a computer program in relation to its sequence of functions (as distinct from the data it processes).

fly•ing mouse /ˈflīiNG ˈmows/ ▸ **n.** a mouse that can be lifted from the desk and used in three dimensions.

fold•er /ˈfōldər/ ▸ **n.** an icon on a computer screen that can be used to access a directory containing related files or documents.

foot•print /ˈfo͝otˌprint/ ▸ **n.** the space taken up on a surface by a piece of computer hardware.

force feed•back /ˈfôrs ˈfēdˌbæk/ ▸ **n.** the simulation of physical attributes such as weight in virtual reality, allowing the user to interact directly with virtual objects using touch: *the device supports force feedback in games* | [as adjective] *a force-feedback steering wheel.*

for•mat /ˈfôrˌmæt/ ▸ **n.** a defined structure for the processing, storage, or display of data: *a data file in binary format.*

▸ **v.** (**for•mat•ted**, **for•mat•ting**) [trans.] (in computing) arrange or put into a format.

■ prepare (a storage medium) to receive data: *format the disk.*

ORIGIN mid 19th cent.: via French and German from Latin *formatus (liber)* 'shaped (book),' past participle of *formare* 'to form.'

form fac•tor /ˈfôrm ˌfæktər/ ▸ **n.** the physical size and shape of a piece of computer hardware.

For•tran /ˈfôrˌtræn/ (also **FORTRAN**) ▸ **n.** a high-level computer programming language used esp. for scientific calculations.

ORIGIN 1950s: contraction of *formula translation.*

FPU ▸ **abbr.** floating-point unit, a processor that performs arithmetic operations.

frag•men•ta•tion /ˌfrægmənˈtāSHən/ ▸ **n.** the storing of a file in parts in many different areas of memory on a hard disk.

frame /frām/ ▸ **n.** a graphic panel in a display window, especially in an Internet browser, that encloses a self-contained section of data and permits multiple independent document viewing: *frames can be a pain, however, in that viewers can no longer use the URL bar to see where they are, and they can't bookmark a frame.*

free•ware /ˈfrē,we(ə)r/ ▸ **n.** software that is available free of charge.

freeze /frēz/ ▸ **v.** (past **froze** /frōz/; past part. **fro•zen** /ˈfrōzən/) [intrans.] (of a computer) become temporarily locked because of a software or system problem: *if a program freezes, you can exit it without shutting down the computer.*

ORIGIN Old English *frēosan* (in the phrase *hit frēoseth* 'it is freezing, it is so cold that water turns to ice'), of Germanic origin; related to Dutch *vriezen* and German *frieren*, from an Indo-European root shared by Latin *pruina* 'hoarfrost' and 'frost.'

front-end /ˈfrənt ˈend/ ▸ **adj.** [attrib.] (of a device or program) directly accessed by the user and allowing access to further devices or programs: *a front-end file server.*

▸ **n.** a part of a computer or program that allows access to other parts.

FTP ▸ **abbr.** file transfer protocol, a standard for the exchange of program and data files across a network.

▸ **v.** (**FTP'd** or **FTPed**, **FTPing**) [trans.] informal transfer (a file) from one computer or system to another using FTP, esp. on the Internet.

full du•plex /ˈfo͝ol ˈd(y)o͞o,pleks/ ▸ **adj.** another term for **DUPLEX**.

func•tion /ˈfəNGkSHən/ ▸ **n.** a basic task of a computer, esp. one that corresponds to a single instruction from the user.

▸ **v.** [intrans.] work or operate in a proper or particular way:

■ (**function as**) fulfill the purpose or task of (a specified thing):

ORIGIN mid 16th cent.: from French *fonction*, from Latin *functio(n-)*, from *fungi* 'perform.'

func•tion•al•i•ty /,fəNGkSHəˈnælit̲ē/ ▸ **n.** the range of operations that can be run on a computer or other electronic system: *new software with additional functionality.*

func•tion key /ˈfəNGkSHən ,kē/ ▸ **n.** a button on a computer keyboard, distinct from the main alphanumeric keys, to which software can assign a particular function.

fuzz•y /ˈfəzē/ ▸ **adj.** (**fuzz•i•er**, **fuzz•i•est**) of or relating to a form of set theory and logic in which predicates may have degrees of applicability, rather than simply being true or false. It has important uses in artificial intelligence and the design of control systems.

THE TEN WORST E-MAIL MISTAKES

Anyone who uses e-mail should avoid the following mistakes:

Giving confidential information in an unsecured e-mail. Your credit card number, for instance, can easily be sent throughout the world. It's best to send credit-card information only through secure Web sites. See "How to Shop Safely Online" on p.143.

Opening attachments from strangers. Never open an e-mail that has an attachment that is vague or says "Check this out!" A virus may spread by invading the contact list on a computer and sending itself to every e-mail address on the list.

Opening unsolicited e-mail without first scanning for viruses. There are several free anti-virus programs available, and they should be updated regularly.

Hitting "reply" to an unsolicited e-mail when asking to be taken off the sender's list. By hitting "reply" you may be opening up your account to a deluge of spam.

Hitting "reply all" when only the sender needs a response. Does everyone really need to know your reply? Think before you reply, especially if the e-mail was sent to a very large group.

Forwarding hoaxes or jokes. Most people get too much e-mail, and they would prefer a real note from you, not a hoary joke or, worse, a scaremongering urban legend or false charity scam. If something sounds too good (or too shocking) to be true, it probably is. Check the web for information before you send something on; www.snopes.com is a great site for checking stories.

Sending an e-mail without a signature. It's helpful to include at least your name and e-mail address at the bottom of your message, especially if you are e-mailing someone for the first time. Don't use a vCard (virtual business card). It may be mistaken for a virus.

Sending an e-mail without spell-checking it. Most e-mail systems spell-check as you type or have a "spell-check before sending" setting.

Sending large files or pictures. Don't clog up your recipient's mailbox—ask before sending big files.

Sending e-mail without a "subject" line or with a vague subject line. Be specific. A subject that reads "Looking forward to dinner Saturday!" will get more attention than one that reads "hi" or "see you soon?" A blank subject line may get no attention at all.

G

game•play /ˈgām,plā/ ▸ **n.** the tactical aspects of a computer game, such as its plot and the way it is played, as distinct from the graphics and sound effects.

gar•bage in, gar•bage out /ˈgärbij ˈin ˈgärbij ˈowt/ (abbreviation: **GIGO** /ˈgī,gō/) ▸ **phrase** used to express the idea that in computing and other spheres, incorrect or poor quality input will always produce faulty output.

ORIGIN late Middle English (in the sense 'offal'): from Anglo-Norman French, of unknown ultimate origin.

gate•keep•ing /ˈgāt,kēpiNG/ ▸ **n.** a function or system that controls access or operations to files, computers, networks, or the like: [as modifier] *you will need to set up a gatekeeping mechanism that allows reads under some circumstances and blocks them under others.*

gate•way /ˈgāt,wā/ ▸ **n.** a device used to connect two different networks, esp. a connection to the Internet.

ga•tor /ˈgātər/ ▸ **v.** (usually **be gatored**) cause a competitor's advertisement to appear on (a commercial Web site): *Gator came into the spotlight mid-2001 for its practice of selling pop-up ads that are delivered to customers visiting rival Web sites, what was then known as getting "Gatored."*

■ cause (a Web site visitor) to view a competitor's advertisement: *he was gatored with an ad for a competitor while visiting his own company's site.*

ORIGIN from *Gator*, the name of the software that creates this effect.

gaze track•ing /ˈgāz ˌtrakiNG/ ▸ **n.** another name for **EYE TRACKING**: *many gaze trackers use a small beam of infrared light reflected off the cornea, which has an outward curve.*

Gbyte /ˈjēˌbīt/ ▸ **abbr.** gigabyte(s).

gen•der bend•er /ˈjendər ˌbendər/ ▸ **n.** a device for changing an electrical or electronic connector from male to female, or from female to male: *we found a nice looking gender bender, took a standard universal Token-Ring workstation cable with a DB-9 male connector and another Token-Ring media filter cable with a DB-9 male connector on one end and twisted-pair RJ-45 jack on the other, used the gender bender in the middle and that was it.*

Gen•er•a•tion D /ˈjenəˈrāSHən ˈdē/ ▸ **n.** the generation of people with great interest or expertise in computers and other digital devices: *their readers are generation D, the digital generation, which adapts very easily to new technology*

ORIGIN from an abbreviation of *digital generation.*

gen•lock /ˈjenˌläk/ ▸ **n.** a device for maintaining synchronization between two different video signals, or between a video signal and a computer or audio signal, enabling video images and computer graphics to be mixed: [as modifier] *if you've got the appropriate genlock device, you can transfer the text to a videotape machine.*

▸ **v.** [intrans.] maintain synchronization between two signals using the genlock technique: *perhaps the only serious criticism of the system regards its ability to genlock.*

ORIGIN 1960s: from *generator* + the verb *lock.*

ge•o•cach•ing /ˈjēōˌkæSHiNG/ ▸ **n.** the recreational activity of hunting for and finding a hidden object by means of Global Positioning System (GPS) coordinates posted on a Web site: *this is the premise of a new sport called geocaching, a 21st-century treasure hunt with a digital spin.*

ORIGIN blend of *geographical* and *cache* 'hide something in a safe place.'

ge•o•mat•ics /ˌjēəˈmatiks/ ▸ **plural n.** [treated as singular] the application of computerization to information in geography and related fields: [as modifier] *a further advantage of geomatics technologies, remote sensing in particular, is that satellites are able to provide* *world-wide coverage on often a continuous basis from spatial resolutions of 1 kilometer down to a few meters.*

DERIVATIVES **ge•o•mat•ic adj.**

ORIGIN 1980s: blend of *geography* and *informatics* 'the science of processing data for storage and retrieval.'

GIF /jif/ ▸ **n.** a popular format for image files, with built-in data compression.

■ (also **gifs**) a file in this format.

ORIGIN late 20th cent.: acronym from *graphic interchange format.*

gig /gig/ ▸ **n.** informal short for **GIGABYTE.**

giga- /ˈgigə/ ▸ **comb. form** used in units of measurement: **1** (of binary data) denoting a factor of 2^{30}.

2 denoting a factor of 10^9: *gigahertz.*

ORIGIN from Greek *gigas* 'giant.'

gig•a•byte /ˈgigəˌbīt/ (abbr.: **GB**) ▸ **n.** a unit of information equal to one billion (10^9) or, strictly, 2^{30} bytes.

gig•a•flop /ˈgigəˌfläp/ ▸ **n.** a unit of computing speed equal to one billion floating-point operations per second.

ORIGIN 1970s: back-formation from *gigaflops* (see **GIGA-**, **-FLOP**).

GIGO /ˈgīˌgō/ ▸ **abbr.** garbage in, garbage out, the idea that incorrect or poor quality input will always produce faulty output.

GIS ▸ **abbr.** geographic information system, a system for storing and manipulating geographical information on computer.

glob•al /ˈglōbəl/ ▸ **adj.** operating or applying through the whole of a file, program, etc: *global searches.*

DERIVATIVES **glob•al•ly adv.**

glyph /glif/ ▸ **n.** a small graphic symbol or character.

DERIVATIVES **glyph•ic adj.**

ORIGIN late 18th cent.: from French *glyphe*, from Greek *gluphē* 'carving.'

goo•gle /ˈgo͞ogəl/ ▸ **trademark** (also **Goo•gle**) informal [intrans.] use an Internet search engine, particularly Google.com: *she spent the afternoon googling aimlessly.* ■ [trans.] search for the name of (someone) on the Internet to find out information about them: *you meet someone, swap numbers, fix a date, then Google them through 1,346,966,000 web pages.*

ORIGIN from *Google*, the proprietary name of a popular Internet search engine.

go•pher /ˈgōfər/ (also **Go•pher**) ▸ **n.** a menu-based system that allows users of the Internet to search for and retrieve documents on topics of interest.

ORIGIN 1990s: named after the gopher mascot of the University of Minnesota, where the system was invented.

grab /græb/ ▸ **n.** [usually with modifier] a frame of video or television footage, digitized and stored as a still image in a computer's memory for subsequent display, printing, or editing: *a screen grab from Wednesday's program.*

graph•i•cal /ˈgræfikəl/ ▸ **adj.** of or relating to computer graphics: *a high-resolution graphical display.*

DERIVATIVES **graph•i•cal•ly adv.**

graph•i•cal us•er in•ter•face /ˈgræfikəl ˈyo͞ozər ˈintərˌfās/ (abbr.: **GUI**) ▸ **n.** a visual way of interacting with a computer using items such as windows, icons, and menus, used by most modern operating systems.

graph•ics /ˈgræfiks/ ▸ **plural n.** (also **com•put•er graph•ics**) [treated as pl.] visual images produced by computer processing.

■ [treated as sing.] the use of computers linked to display screens to generate and manipulate visual images. ■ (often **graphic**) an input or computer-generated image that can be displayed on a screen and manipulated with software tools.

ORIGIN mid 17th cent.: via Latin from Greek *graphikos*, from *graphē* 'writing, drawing.'

graph•ics card /ˈgræfiks ˌkärd/ ▸ **n.** a printed circuit board that controls the output to a display screen.

graph•ics tab•let /ˈgræfiks ˌtæblit/ ▸ **n.** an input device consisting of a flat, pressure-sensitive pad that the user draws on or points at with a special stylus, to guide a pointer displayed on the screen.

gray /grā/ (British **grey**) ▸ **v.** [trans.] (**gray something out**) display a menu option in a light font to indicate that it is not available: [usually passive] *when I right-click on the icons, all the property fields on the Shortcut tab are either missing or grayed out.*

gray•scale /ˈgrāˌskāl/ ▸ **n.** a range of gray shades from white to black, as used in a monochrome printout: [as adjective] *a grayscale scanner.*

grid /grid/ ▸ **n.** a number of computers linked together via the Internet so that their combined power may be harnessed to work on difficult problems.

group•ware /ˈgro͞op,we(ə)r/ ▸ **n.** software designed to facilitate collective working by a number of different users.

GUI /ˈgo͞oē/ ▸ **abbr.** graphical user interface.

H

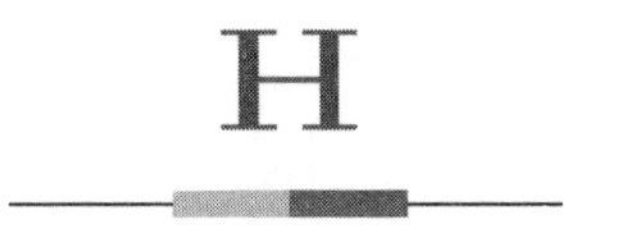

hack /hæk/ ▸ **v.** [trans.] skillfully program (a computer) or revise (program code): *hack a driver for a printer.*
■ [intrans.] use a computer to gain unauthorized access to data in a system: *they* ***hacked into*** *a bank's computer.* ■ [trans.] gain unauthorized access to (data in a computer): *hacking private information from computers.*
▸ **n.** informal an act of computer hacking.
■ a piece of computer code that performs some function, esp. an unofficial alternative or addition to a commercial program: *freeware and shareware hacks.*
ORIGIN Old English *haccian* 'cut in pieces,' of West Germanic origin; related to Dutch *hakken* and German *hacken.*

hack•er /ˈhækər/ ▸ **n.** informal an enthusiastic and skillful computer programmer or user.
■ a person who uses computers to gain unauthorized access to data.

half-du•plex /ˈhæf ˈd(y)o͞oˌpleks/ ▸ **adj.** (of a communications system or computer circuit) allowing the transmission of signals in both directions but not simultaneously: *the network is half-duplex.* Also see **DUPLEX.**

hand•held /ˈhændˌheld/ (also **hand•held com•put•er**) ▸ **n.** a small lightweight computer that can be easily held in one hand. Also see **PDA.**

hand•shake /ˈhæn(d)ˌSHāk/ ▸ **n.** an exchange of standardized signals between devices in a computer network regulating the transfer of data.
DERIVATIVES **hand•shak•ing n.**

hands-on /ˈhæn(d)z ˈôn/ ▸ **adj.** involving or requiring personal operation at a keyboard.

■ involving or offering active participation rather than theory: *hands-on computer programming experience.*

hang /hæNG/ ▸ **v.** (past and past part. **hung** /həNG/) come or cause to come unexpectedly to a state in which no further operations can be carried out: *the program would consistently hang.*

ORIGIN Old English *hangian* (intransitive verb), of West Germanic origin, related to Dutch and German *hangen*, reinforced by the Old Norse transitive verb *hanga.*

hap•tic /ˈhæptik/ ▸ **adj.** of or relating to the sense of touch through a computer interface and the perception and manipulation of objects in virtual reality: *haptic devices can be an aid to computer users who are visually impaired.* Also see **FORCE FEEDBACK.**

ORIGIN late 19th cent.: from Greek *haptikos* 'able to touch or grasp,' from *haptein* 'fasten.'

hard-code /ˈhärd ˈkōd/ ▸ **v.** [trans.] fix (data or parameters) in a program in such a way that they cannot easily be altered by the user.

hard cop•y /ˈhärd ˈkäpē/ ▸ **n.** a printed version on paper of data held in a computer.

hard disk /ˈhärd ˈdisk/ ▸ **n.** a rigid nonremovable magnetic disk with a large data storage capacity, as distinct from the smaller capacity floppy disk.

hard drive /ˈhärd ˈdrīv/ ▸ **n.** a high-capacity, self-contained storage device containing a read-write mechanism plus one or more hard disks, inside a sealed unit. (Also called **hard disk drive.**)

hard er•ror /ˈhärd ˈerər/ ▸ **n.** an error or hardware fault causing failure of a program or operating system, esp. one that gives no option of recovery.

hard par•ti•tion•er /ˈhärd pärˈtiSHənər/ ▸ see **PARTITIONER.**

hard•ware /ˈhärdˌwe(ə)r/ ▸ **n.** the machines, wiring, and other physical components of a computer or other electronic system. Compare with **SOFTWARE.**

hard•wir•ing /ˈhärdˈwīriNG/ ▸ **n.** electronic connection by means of wires: *to achieve effective automation of a line of this type, Eurobend would have had to use numerous handshaked PLC units* *and extensive hardwiring, as well as spending many hours on system programming.*

HDD ▸ **abbr.** hard disk drive.

head /hed/ ▸ **n.** a component in a computer drive by which data is transferred to and from a magnetic disk or tape.

■ short for **PRINTHEAD.**

ORIGIN Old English *hēafod*, of Germanic origin; related to Dutch *hoofd* and German *Haupt*.

help /help/ ▸ **n.** a set of searchable information and instructions for a computer or software program: [as adjective] *a help menu.*

ORIGIN Old English *helpan* (verb), *help* (noun), of Germanic origin; related to Dutch *helpen* and German *helfen*.

help desk /'help ˌdesk/ ▸ **n.** a service providing information and support to the users of a computer network or product: *when you work on the help desk, you see every dumb thing people can do to and with their computers.*

heu•ris•tic /hyo͞o'ristik/ ▸ **adj.** proceeding to a solution by trial and error or by rules that are only loosely defined: *this program includes built-in heuristic anti-spam technology.*

▸ **n.** a heuristic process or method.

DERIVATIVES **heu•ris•ti•cal•ly adv.**

ORIGIN early 19th cent.: formed irregularly from Greek *heuriskein* 'find.'

hex /heks/ ▸ **adj.** short for **HEXADECIMAL.**

hex•a•dec•i•mal /ˌheksə'des(ə)məl/ ▸ **adj.** relating to or using a system of numerical notation that has 16 rather than 10 as its base: *hexadecimal values.*

DERIVATIVES **hex•a•dec•i•mal•ly adv.**

hi•ber•nate /'hībərˌnāt/ ▸ **v.** [intrans.] automatically save data in RAM to the hard drive and turn off the monitor and computer to reduce power use, usually after a computer has been unused for a specified period of time.

DERIVATIVES **hi•ber•na•tion** /ˌhībər'nāsʜən/ **n.**

ORIGIN early 19th cent. from Latin *hibernare*, from *hiberna* 'winter quarters,' from *hibernus* 'wintry.'

hi•er•ar•chy /'hī(ə)ˌrärkē/ ▸ **n.** (pl. **hi•er•ar•chies**) an arrangement or classification of things according to relative importance or inclusiveness: *their software allows users to create and maintain a sophisticated filing hierarchy to manage their records.*

DERIVATIVES **hi•er•ar•chize** /–ˌkīz/ **v.**

ORIGIN: late Middle English: via Old French and medieval Latin from Greek *hierarkhia*, from *hierarkhēs* 'sacred ruler.' The earliest sense was 'system of orders of angels and heavenly beings.'

high-lev•el /ˈhī ˌlevəl/ ▸ **adj.** denoting a programming language (e.g., BASIC or Java) that is relatively accessible to the user, having instructions that resemble an existing language such as English. Compare with **LOW-LEVEL**.

hit /hit/ ▸ **n. 1** an instance of a particular Web site being accessed by a user: *the site gets an average 350,000 hits a day—that's a staggering four per second.*

2 a single instance of finding a search target, on the Internet or in a database: *It's not fancy, and sometimes browsing gets you more relevant hits, but it is easy.*

home page /ˈhōm ˌpāj/ (also **home•page**) ▸ **n.** the introductory document of an individual's or organization's Web site. It typically serves as a table of contents to the site's other pages or provides links to other sites.

host /hōst/ ▸ **n.** (also **host com•put•er** /ˈhōst kəmˌpyo͞otər/) a computer that mediates multiple access to databases mounted on it or provides other services to a computer network.

ORIGIN Middle English: from Old French *hoste*, from Latin *hospes*, *hospit-* 'host, guest.'

hot key /ˈhät ˌkē/ ▸ **n.** a key or a combination of keys providing quick access to a particular function within a program.

hot link /ˈhät ˌliNGk/ ▸ **n.** a connection between documents or applications that enables material from one source to be incorporated into another; in particular, a facility that automatically updates material in a document when an alteration is made to the document from which it originated.

■ a hypertext link.

▸ **v.** (**hot-link** /ˈhät ˈliNGk/) [trans.] connect (two documents) by means of a hot link.

hot•list /ˈhätˌlist/ ▸ **n.** a personal list of favorite or most frequently accessed Web sites compiled by an Internet user.

hot spot /ˈhät ˌspät/ ▸ **n. 1** an area on the screen that can be clicked on to start an operation such as loading a file.

2 a place where a wireless network providing connection to the Internet is operating and can be accessed by a wireless-equipped computer: *the airport terminals are hot spots* | [as adjective] *hot spot locations.*

hot-swap /ˈhät ˈswäp/ ▸ **v.** [trans.] informal fit or replace (a computer part) with the power still connected.

DERIVATIVES **hot-swap•pa•ble adj.**

house•keep•ing /ˈhowsˌkēpiNG/ ▸ **n.** operations such as record-keeping or maintenance in an organization or a computer that make work possible but do not directly constitute its performance.

HTML ▸ **n.** hypertext markup language, a standardized system for tagging text files to achieve font, color, graphic, and hyperlink effects on World Wide Web pages.

HTTP ▸ **abbr.** hypertext transfer (or transport) protocol, the data transfer protocol used on the World Wide Web.

hy•per•link /ˈhīpərˌliNGk/ ▸ **n.** a link from a hypertext file or document to another location or file, typically activated by clicking on a highlighted word or image on the screen.

▸ **v.** [trans.] link (a file) in this way: *thumbnail images that are hyperlinked to a larger image.*

hy•per•text /ˈhīpərˌtekst/ ▸ **n.** a software system that links topics on the screen to related information and graphics, which are typically accessed by a point-and-click method.

■ a document presented on a computer in this way.

BUYING A COMPUTER

A previous essay (see "Which Operating System Is Right for You," p. 16), pointed out that your choice of hardware depends largely on your choice of operating system. So let's assume that you've made up your mind and decided whether you want a Mac or a PC. The next question is whether you want a *desktop* or a *laptop*. Ask yourself a couple of relevant questions. Do you work in one place or you are a "road warrier"? Is space a consideration? Will this be your only computer? And how much can you afford?

Desktops

Feature for feature, desktops are cheaper than laptops. Normally, they are also more expandable. You can add more RAM, a better video card, an additional hard drive, a DVD or CD burner, or any of a number of other devices—provided that there are empty slots on the motherboard for additional expansion boards and empty drive bays in the case. You might eventually be able to add a more powerful processor, thus prolonging the useful life of your computer. Desktop cases are relatively easy to open, making it possible for intrepid users to install these items themselves. So check for expandability before you buy.

You'll need a monitor, of course. You'll pay a bit less for one that comes with your desktop computer, but the dread sign in computer ads saying, "Monitor sold separately," may not be such a bad thing. You may want something better than your computer vendor is offering. Having a large, steady, clear monitor can make a real difference in the quality of your computing experience. Broadly, there are two kinds to choose from now, and both flat-panel monitors and CRTs have distinct advantages.

Sales of *flat-panel monitors* are fast catching up with those of *CRTs*, but—as always—your choice will depend on how you plan to use your computer, how much space you have, and the extent to which you are seduced by beauty and coolness. Flat-panel monitors, with their clean, slender profiles, are indeed appealing. (To find out why, see "The Ten Best Tools and Peripherals You Didn't Know About," p. 196.) New flat panels (and, by the way, laptop screens) come with *TFT* displays (that is, *active-matrix,* not the older, fuzzier *passive-matrix* displays), and their glare-free screens are easy on the eyes. In Windows XP, you can make the text on an LCD screen even sharper by choosing ClearType (Display >Properties >Appearance >Effects). A nice little freeware program for fine-tuning ClearType can be found at http://www.ioisland.com/cleartweak(.) ClearType does not work well, however, on a CRT.

If you do choose a flat-panel monitor, you may want to check the manufacturer's "dead pixel" policy. They may require a minimum number of unlit pixels (perhaps five) before they're willing to replace the monitor. Fortunately, dead pixels are fairly rare.

CRTs, which have a faster response time, remain the choice for gamers and are still preferred by graphics professionals for their fine color reproduction. They are also less expensive than flat panels. On either type of monitor, look for dot pitch specifications of .26 to .28 millimeters, find out if you need an adapter to connect it to your computer, and check to see if your video card is up to snuff.

Laptops

If you can afford the price premium on a laptop, it offers many advantages. In a small office or apartment, a laptop that can be stowed away in a drawer or on a shelf is a great space saver. You can work in any room, take your computer to the library to do research, play DVDs on airplanes, access e-mail from remote locations, and take work with you on vacation. (Wait! Maybe that's not an advantage!)

But there are downsides. If you're traveling (and especially if you're somewhat harried), you can lose your laptop in a taxi; you may not be able to find an outlet when the battery runs out—or a phone line—or a wireless connection to the Internet; and managing the adapter, external drives, and various tangled cords and cables is not easy to do on the road. In addition, laptops are vulnerable. They don't take well to being dropped, hit, or yanked off a desk by the cord, and they are irresistible to thieves. You'll need something like a Kensington lock to chain the computer to furniture in public places (see http://www.microsaver.com).

Although you can't easily add new hardware (or even memory) inside most laptops—at least not yourself—they almost always include USB ports, Firewire ports, and PC Card slots. These allow you to attach a wide variety of peripherals (e.g., scanners, printers, digital cameras, and CD-ROM drives), as well as providing the option of network and online interfaces. Of course the more peripherals you buy, the more things you have to carry around.

As a matter of fact, weight is something you should consider carefully before you make your initial purchase. So-called desktop replacement laptops can weigh from seven to ten or more pounds. And when you're racing through the airport, they apparently gain weight with every step. For the mobile but fairly strong user, there are laptops in the four- to six-plus-pound range. If weight is a real issue, and you'll be carrying the computer frequently, "thin-and-light" laptops are available, some weighing less than three pounds! A laptop may well be worth the cost, but in making your decision, try not to overestimate your need for mobility. Make sure you're spending the extra money for the right reasons.

If you do succumb, a good, well-padded laptop bag is a necessity. Avoid one with a shape that screams, "I'm a computer case"! It will just attract thieves. And if your laptop weighs more than six pounds, you may want to consider a backpack or a rolling computer bag rather than a disguised briefcase.

How much computer power?
New small-business and home computers range in cost from, roughly, a shocking $400 to an equally shocking $3,000 plus. Or more. If you know what you'll be using your computer for, you have a good idea of where you are in that range; you obviously have to pay for more processor speed and the latest technology. If your computing will be limited to e-mail and the Internet, plus occasional word processing, a low-end computer (actually more powerful than the expensive high-end computers of, say, two years ago!) will more than suffice. But gamers need high-end graphics and speed. And if you're a college student or you work at home, you're virtually guaranteed an eventual need for functionality that you can't even anticipate now. In a word, if you're a potential power user, you should probably buy as much computer power as you can afford.

Where to buy?
In general, consumers report a high degree of satisfaction in dealing with well-known online retailers. While you may like the instant gratification of buying in a local computer store, you are unlikely to find the best deals, choices, or expertise there. One tactic is to check out floor models in your bricks-and-mortar retailer and then order what you like online. Many online vendors run constant promotions, tempting you with anything from extra RAM or hard drive upgrades to CD burners. Check your vendor to see if you will be subject to a restocking fee in the event that you must return the machine.

Unless you're extremely computer-savvy, buying a computer from an online auction house, a classified ad, a garage sale, or someone who will make it out of miscellaneous parts is not recommended. You risk not having the backing of a warranty or professional service should something go wrong. But wherever you buy—large computer store, giant warehouse, or online vendor—do some research first. Exploring physical stores and online sites, reading a couple of issues of a good computer magazine, and asking your knowledgeable friends for advice will pay off in the end.

Warranty or No Warranty? Service Plan or No Service Plan?

Most new computers come with a certain amount of free technical support, starting from either the date of purchase or the date of registration (be sure you know which if you are buying a computer that you will not be setting up immediately, e.g., one to take to college). However, keep in mind that many of these "free" support plans are hard to use—you may spend a lot of time listening to bad music while on hold. Other plans are online only, through e-mail or chat—hard to take advantage of if your computer is not working properly! Always write down any error messages you receive—word for word and number for number. You'll need that information when you call technical support.

Warranties vary in what they cover and for how long. Make sure you understand the terms of yours before you buy. Plans that allow for house calls by a technician are vastly superior to those that require you to ship your computer to a central repair facility or take it to an authorized dealer, but they are usually more expensive. If you don't have computer-knowledgeable friends or relatives to help you, or a local dealer or repair center nearby, it may be worth buying an extended service plan or warranty, at least for the first year. After a year or two, if your computer is working well, you can wait until the lure of new technology overpowers you. If, on the other hand, it has turned into a lemon, it may be less trouble to replace the entire machine (with one from a different manufacturer!) than to try to revive it further.

Opening the Box

When you get your new computer, set aside a block of uninterrupted time to set it up. Setting a up a new computer late at night is not a good idea. You're tired, and support lines are likely to be unavailable. Keep all the paperwork and manuals, as well as any accessories or cords you don't have an immediate need for, in a safe place. Ask one of your computer-savvy friends to help. (Treat them to pizza!) Get them to explain what they are doing, and which cord goes where, so that you can duplicate the process if necessary. If you are installing software that was installed on your previous computer, don't dismantle that computer until you are sure you have all the registration numbers for your software! (You can usually find them by looking for "about [program name]" under the Help menu.)

If you have the space, keep the boxes for a week or two. You can toss them once you're sure your computer is working properly and you won't have to send it back.

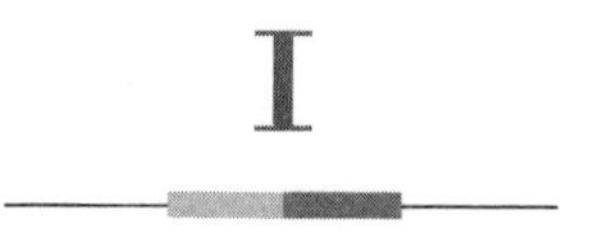

IC ▸ **abbr.** ■ integrated circuit.

ICANN ▸ **abbr.** Internet Committee for Assigned Names and Numbers, the nonprofit organization that oversees the use of Internet domains.

i•con /ˈīˌkän/ ▸ **n.** a symbol or graphic representation on a computer screen of a program, option, or window, esp. one of several for selection.

ORIGIN mid 16th cent.: via Latin from Greek *eikōn* 'likeness, image.'

i•con•i•fy /īˈkänəˌfī/ ▸ **v.** (**i•con•i•fies, i•con•i•fied**) [trans.] **ICONIZE.**

i•con•ize /ˈīkəˌnīz/ ▸ **v.** [trans.] reduce (a window on a video display terminal) to a small symbol or graphic: *unlike Windows, which right out of the box iconizes background programs and puts them in a line across the bottom of the screen.*

ICT ▸ **abbr.** information and computing technology.

IDE ▸ **abbr.** integrated drive electronics, a standard used primarily for connecting disk drives to a computer.

i•den•ti•fi•er /īˈdentəˌfīər/ ▸ **n.** a sequence of characters used to identify or refer to a program or an element, such as a variable or a set of data, within it.

IEEE 1394 (also **FireWire**) ▸ **n.** a standard for high-speed connection and data exchange between computers and peripherals: *both front and rear IEEE 1394 ports.*

ORIGIN 1990s an abbreviation of *Institute of Electrical and Electronics Engineers.*

IEEE 802.11 ▸ **n.** a standard for communication over a wireless network. See also **WI-FI.**

ORIGIN 1990s an abbreviation of *Institute of Electrical and Electronics Engineers.*

IM ▸ **abbr.** ■ instant message. ■ instant messaging.

▸**v.** [trans.] (**IM's, IM'd, IM'•ing**) send a message to (someone) by using an instant messaging system: *by now I was being IM'd (instant messaged) by a tireless horde of hot-blooded all-American testosterone-crazed males.*

im•age /ˈimij/ ▸ **v.** [trans.] (usu. **be imaged**) make a visual representation of (something) by scanning it with a detector or electromagnetic beam: *every point on the Earth's surface was imaged by the satellite* | [as noun] (**imaging**) *medical imaging.*

ORIGIN Middle English: from Old French, from Latin *imago.*

IMAP /ˈīˌmæp/ ▸ **abbr.** Internet Mail Access Protocol.

im•bed /imˈbed/ ▸ see **EMBED**.

im•mer•sive /iˈmərsiv/ ▸ **adj.** (of a computer display or system) generating a three-dimensional image that appears to surround the user.

i-Mode /ˈī ˌmōd/ ▸ **trademark** a technology that allows data to be transferred to and from Internet sites via cell phones: *i-Mode is the internet-on-your-mobile service you see kids in Tokyo using on those funky flip-tops with the palm-size screens.*

ORIGIN early 21st cent.: from *I* (referring to the user's ability to interact directly with the Internet) + *mode.*

im•port /imˈpôrt/ ▸ **v.** [trans.] transfer (data) into a file or document. ■ transfer (digital files) to a computer from another device: *software that imports pictures from camera to computer.*

DERIVATIVES **im•port•a•ble adj.**; **im•por•ta•tion** /ˌimpôrˈtāsHən/ **n.**

ORIGIN from Latin *importare* 'bring in' (in medieval Latin 'imply, mean, be of consequence'), from *in-* 'in' + *portare* 'carry.'

im•pres•sion /imˈpresHən/ ▸ **n.** an instance of a pop-up or other Web advertisement being seen on computer users' monitors: *in July, Nielsen/NetRatings reported Web publishers served 7.3 billion pop-up ad impressions.*

in-box /ˈin ˌbäks/ ▸ **n.** the window in which an individual user's received e-mail messages and similar electronic communications are displayed: *another version from Puma uses Orchestrate, a Web-based universal in-box for faxes, e-mail and voice mail.*

in•cre•men•tal back•up /ˈiNGkrəˈmentl ˈbækˌəp/ ▸ **n.** a security copy that contains only those files that have been altered since the last full backup.

in•dex /ˈinˌdeks/ ▸ **n.** (pl. **in•dex•es** or esp. in technical use **in•di•ces** /ˈindəˌsēz/) a set of items each of which specifies one of the records of a file and contains information about its address.

DERIVATIVES **in•dex•a•ble**, **in•dex•i•ble** adj.; **in•dex•a•tion** /ˌindekˈsāSHən/ n.; **in•dex•er** n.

ORIGIN late Middle English: from Latin *index*, *indic-* 'forefinger, informer, sign,' from *in-* 'toward' + a second element related to *dicere* 'say' or *dicare* 'make known'. The original sense 'index finger' (with which one points) came to mean 'pointer' (late 16th cent.), and figuratively something that serves to point to a fact or conclusion; hence a list of items ("pointing" to their location).

in•fect /inˈfekt/ ▸ **v.** [trans.] affect with a virus.

DERIVATIVES **in•fec•tor** **n.**

ORIGIN late Middle English: from Latin *infect-* 'tainted,' from the verb *inficere*, from *in-* 'into' + *facere* 'put, do.'

in•fec•tion /inˈfekSHən/ ▸ **n.** the presence of a virus in, or its introduction into, a computer system.

ORIGIN late Middle English: from late Latin *infectio(n-)*, from Latin *inficere* 'dip in, taint' (see **INFECT**).

in•fo•me•di•ar•y /ˌinfōˈmēdēˌerē/ ▸ **n.** an Internet company that gathers and links information on particular subjects on behalf of commercial organizations and their potential customers: *"Infomediaries sit between buyers and sellers," says the managing director for an investment bank. "They market everything from blueberries to backhoes."*

ORIGIN 1980s: from *info(rmation)* + *-mediary*, on the pattern of *intermediary*.

in•for•mat•ics /ˌinfərˈmætiks/ ▸ **plural n.** [treated as sing.] the science of processing data for storage and retrieval; information science.

ORIGIN 1960s: from **INFORMATION** + *-ics*, translating Russian *informatika*.

in•for•ma•tion /ˌinfərˈmāSHən/ ▸ **n.** data as processed, stored, or transmitted by a computer.

■ (in information theory) a mathematical quantity expressing the probability of occurrence of a particular sequence of symbols, impulses, etc., as contrasted with that of alternative sequences.
DERIVATIVES **in•for•ma•tion•al adj.**
ORIGIN late Middle English, via Old French from Latin *informatio(n-)*, from the verb *informare*.

in•for•ma•tion re•triev•al /ˌinfərˈmāsнən riˌtrēvəl/ ▸ **n.** the tracing and recovery of specific information from stored data.

in•forma•tion scent /ˌinfərˈmāsнən ˌsent/ ▸ **n.** visual or textual cues provided on a Web site to suggest what information it or its links may contain: *not unlike an animal's foraging behavior users use "information scent" to optimize their efforts to find what they want.*
■ the perceived usefulness of a page based on such information.

in•for•ma•tion sci•ence /ˌinfərˈmāsнən ˌsīəns/ ▸ **n.** the study of processes for storing and retrieving information, esp. scientific or technical information.

in•for•ma•tion su•per•high•way /ˈinfərˈmāsнən ˈso͞opərˈhīˌwā/ ▸ **n.** an extensive electronic communications network such as the Internet, used for the rapid transfer of information such as sound, video, and graphics in digital form.

in•for•ma•tion tech•nol•o•gy /ˌinfərˈmāsнən tekˌnäləjē/ (abbr.: **IT**) ▸ **n.** the study or use of systems (esp. computers and telecommunications) for storing, retrieving, and sending information.

in•fo•war /ˈinfōˌwôr/ ▸ **n. 1** hostile actions against an enemy's information infrastructure: *Yugoslav hackers are waging infowar against NATO's computer network, but the fighters appear to be poorly armed.*
2 a propaganda war waged via electronic media.

in•i•tial•ize /iˈnisнəˌlīz/ ▸ **v.** [trans.] **1** (often **be initialized to**) set to the value or put in the condition appropriate to the start of an operation: *the counter is initialized to one.*
2 format (a computer disk).
DERIVATIVES **in•i•tial•i•za•tion n.**

ink-jet print•er /ˈiNGk ˌjet ˈprintər/ (also **ink•jet print•er**) ▸ **n.** a printer in which the characters are formed by minute jets of ink.

in-line /ˈin ˈlīn/ ▸ **adj.** constituting an integral part of a computer program: *the parameters can be set up as in-line code.*
■ constituting an integral part of a continuous sequence of operations or machines.

in•put /ˈinˌpo͝ot/ ▸ **n.** the information fed into a computer or computer program: *pen-based computers take input from a stylus.*
■ the action or process of putting or feeding information in: *the input of data to the system.* ■ energy supplied to a device or system; an electrical signal: *the input is a low-frequency signal.*
▸ **v.** (**in•put•ting**; past and past part. **input** or **in•put•ted**) [trans.] put (data) into a computer.
DERIVATIVES **in•put•ter n.**

in•stant mes•sag•ing /ˈinstənt ˈmesijiNG/ (abbreviation: **IM**) ▸ **n.** the exchange of typed messages between computer users in real time via the Internet: *AOL heavily promotes its features such as chat room, instant messaging, and (filtered) Web access*
DERIVATIVES **in•stant mes•sage n.**

in•struc•tion /inˈstrəkSHən/ ▸ **n.** (often **instructions**) a code or sequence in a computer program that defines an operation and puts it into effect.
ORIGIN late Middle English: via Old French from late Latin *instructio(n-)*, from the verb *instruere.*

in•struc•tion set /inˈstrəkSHən ˌset/ ▸ **n.** the complete set of all the instructions in machine code that can be recognized and executed by a central processing unit.

in•te•grat•ed cir•cuit /ˈintiˌgrātid ˈsərkət/ (abbr.: **IC**) ▸ **n.** an electronic circuit etched on a small piece of semiconducting material, usually silicon.

in•teg•ri•ty /inˈtegritē/ ▸ **n.** internal consistency or lack of corruption in electronic data: [as adjective] *integrity checking.*
ORIGIN late Middle English: from French *intégrité* or Latin *integritas*, from *integer* 'intact'.

in•tel•li•gent /inˈtelijənt/ ▸ **adj.** (esp. of a computer terminal) incorporating a microprocessor and having its own processing capability. Often contrasted with **DUMB**.
ORIGIN early 16th cent.: from Latin *intelligent-* 'understanding,'

from the verb *intelligere*, variant of *intellegere* 'understand,' from *inter* 'between' + *legere* 'choose.'

in•ter•ac•tive /ˌintərˈæktiv/ ▸ **adj.** (of a computer or other electronic device) allowing a two-way flow of information between it and a user, responding to the user's input: *interactive video.*

DERIVATIVES **in•ter•ac•tive•ly adv.**; **in•ter•ac•tiv•i•ty** /-ækˈtivitē/ **n.**

in•ter•face /ˈintərˌfās/ ▸ **n.** a device or program enabling a user to communicate with a computer.

■ a device or program for connecting two items of hardware or software so that they can be operated jointly or communicate with each other.

▸ **v.** [intrans.] (**interface with**) connect with (another computer or piece of equipment) by an interface.

USAGE: The word **interface** is a relatively new word, having been in the language (as a noun) since the 1880s. In the 1960s it became widespread in computer use and, by analogy, began to be used as both a noun and a verb in many other spheres. Traditionalists object to it on the grounds that there are plenty of other words that are more exact and sound less like "vogue jargon." The verb **interface** is best restricted to technical references to computer systems, etc.

in•ter•leave /ˌintərˈlēv/ ▸ **v.** [trans.] divide (memory or processing power) between a number of tasks by allocating successive segments of it to each task in turn.

■ mix (two or more digital signals) by alternating between them.

In•ter•net /ˈintərˌnet/ ▸ **n.** an international computer network providing e-mail and information from computers in educational institutions, government agencies, and industry, accessible to the general public via modem links.

ORIGIN late 20th cent.: from *inter-* 'reciprocal, mutual' + **NETWORK**.

In•ter•net serv•ice pro•vid•er /ˈintərˌnet ˈsərvis prəˌvīdər/ (abbr. **ISP**) (also **In•ter•net ac•cess pro•vid•er** /ˈintərˌnet ˈækses prə ˌvīdər/) ▸ **n.** see **SERVICE PROVIDER**.

in•ter•op•er•a•bil•i•ty /ˌintərˌäp(ə)rəˈbilitē/ ▸ **n.** the ability of two or more systems with different architecture, platforms, or the like to share information: *in many buildings, interoperability means that* *the front-end controls for the various systems in the building all terminate in the same room.*

in•ter•op•er•a•ble /ˌintər'äp(ə)rəbəl/ ▸ **adj.** (of computer systems or software) able to exchange and make use of information.

DERIVATIVES **in•ter•op•er•a•bil•i•ty** /-ˌäp(ə)rə'bilit̲ē/ **n.**

in•ter•pret•er /in'tərprit̲ər/ ▸ **n.** a program that can analyze and execute a program line by line.

ORIGIN late Middle English: from Old French *interpreteur*, from late Latin *interpretator*, from Latin *interpretari*.

in•ter•ro•gate /in'terəˌgāt/ ▸ **v.** [trans.] obtain data from (a computer file, database, storage device, or terminal).

■ (of an electronic device) transmit a signal to (another device) to obtain a response giving information about identity, condition, etc.

ORIGIN late 15th cent.: from Latin *interrogat-* 'questioned,' from the verb *interrogare*, from *inter-* 'between' + *rogare* 'ask.'

in•ter•work /ˌintər'wərk/ ▸ **v.** [intrans.] (of items of hardware or software) be able to connect, communicate, or exchange data: *servers running new and old versions of the software will interwork.*

in•tra•net /'intrəˌnet/ (also **In•tra•net**) ▸ **n.** a local or restricted communications network, especially a private network accessible via ordinary Internet connectivity software: *intranets may be used entirely in isolation from the outside world. . .or, as is more likely given the benefits of doing so, they may be hooked up to the global Internet.*

in•tu•i•tive /in't(y)o͞oit̲iv/ ▸ **adj.** (chiefly of computer software) easy to use and understand: *an intuitive user interface.*

DERIVATIVES **in•tu•i•tive•ly adv.**

ORIGIN late 15th cent. (originally used of sight, in the sense 'accurate, unerring'): from medieval Latin *intuitivus*, from Latin *intueri*.

in•val•id /in'vælid/ ▸ **adj.** (of computer instructions, data, etc.) not conforming to the correct format or specifications.

DERIVATIVES **in•val•id•ly adv.**

ORIGIN mid 16th cent. (earlier than *valid*): from Latin *invalidus*, from *in-* 'not' + *validus* 'strong'.

IP ▸ **abbr.** Internet protocol, the method by which information is sent between any two Internet computers on the Internet.

IP ad•dress /'ī'pē ə,dres; ,ædris/ ▸ **n.** a unique string of numbers separated by periods that identifies each computer attached to the Internet. It also usually has a version containing words separated by periods: *web server logs use host names or IP addresses to tell where users are coming from.*

ORIGIN from the abbreviation of *Internet protocol.*

IRC ▸ **abbr.** Internet relay chat, an area of the network where users can communicate interactively with each other.

ISA ▸ **abbr.** industry standard architecture, a standard for connecting computers and their peripherals: [as adj.] *an ISA expansion slot.*

ISDN ▸ **abbr.** integrated services digital network, a telecommunications network through which sound, images, and data can be transmitted as digitized signals.

i•soch•ro•nous /ī'säkrənəs/ ▸ **adj.** (of data transmission) requiring an equal rate of data flow: *isochronous audio and video.*

DERIVATIVES **i•soch•ro•nous•ly adv.**

ORIGIN early 18th cent.: from modern Latin *isochronus* (from Greek *isokhronos*, from *isos* 'equal' + *khronos* 'time') + *-ous.*

ISP ▸ **abbr.** Internet service provider: *the fear is that ISPs will not be able to keep up with the rate of growth in traffic on their respective networks or will not be able to muster sufficient revenues and profit margins to justify further investment to expand their capacity.*

ISV ▸ **abbr.** independent software vendor.

IT ▸ **abbr.** information technology.

it•er•ate /'it̲ə,rāt/ ▸ **v.** [intrans.] make repeated use of a mathematical or computational procedure, applying it each time to the result of the previous application; perform iteration.

ORIGIN mid 16th cent.: from Latin *iterat-* 'repeated,' from the verb *iterare*, from *iterum* 'again.'

it•er•a•tion /,it̲ə'rāsHən/ ▸ **n. 1** a new version of a piece of computer hardware or software.

2 repetition of a mathematical or computational procedure applied to the result of a previous application, typically as a means of obtaining successively closer approximations to the solution of a problem.

ORIGIN late Middle English: from Latin *iteratio(n-)*, from the verb *iterare* (see **ITERATE**).

it•er•a•tive /ˈit̲əˌrāt̲iv/ ▸ **adj.** relating to or involving iteration, esp. of a mathematical or computational process.

DERIVATIVES **it•er•a•tive•ly adv.**

ORIGIN late 15th cent.: from French *itératif*, *-ive*, from Latin *iterare* 'to repeat'.

J

jack /jæk/ ▸**n.** a socket with two or more pairs of terminals, designed to receive a jack plug.

▸**jack in** (or **into**) informal log into or connect up (a computer or electronic device).

Ja•va /ˈjävə/ ▸**trademark** a general-purpose computer programming language designed to produce programs that will run on any computer system.

JCL ▸**abbr.** job control language.

job /jäb/ ▸**n.** an operation or group of operations treated as a single and distinct unit.

job con•trol lan•guage /ˈjäb kənˌtrōl ˌlæNGgwij/ ▸**n.** a language enabling the user to define the tasks to be undertaken by the operating system.

joy•pad /ˈjoiˌpæd/ ▸**n.** an input device for a computer game that uses buttons to control the motion of an image on the screen.

ORIGIN late 20th cent.: blend of *joystick* and *keypad.*

joy•stick /ˈjoiˌstik/ ▸**n.** informal a lever that can be moved in several directions to control the movement of an image on a computer or similar display screen.

JPEG /ˈjāˌpeg/ ▸**n.** a format for compressing images: [as adjective] *a JPEG image.*

ORIGIN 1990s: abbreviation of *Joint Photographic Experts Group.*

juke•box /ˈjo͞okˌbäks/ ▸**n.** a device that stores several computer disks in such a way that data can be read from any of them.

ORIGIN 1930s: from Gullah *juke* 'disorderly' + **BOX**[1].

jump in•struc•tion /ˈjəmp inˌstrəkSHən/ ▸**n.** an instruction in a computer program that causes processing to move to a different place in the program sequence.

jump•sta•tion /ˈjəmp ˌstāsʜən/ ▸ **n.** a site on the World Wide Web containing a collection of hypertext links, usually to pages on a particular topic: *use the form below to search the GIS Jump Station.*

K

K ▸**abbr.** kilobyte(s).

KB ▸**abbr.** (also **Kb**) kilobyte(s).

Kbps ▸**abbr.** kilobits per second.

kbyte /ˈkāˌbīt/ ▸**abbr.** kilobyte(s).

ker•nel /ˈkərnl/ ▸**n.** the most basic level or core of an operating system of a computer, responsible for resource allocation, file management, and security.

key /kē/ ▸**n.** (pl. **keys**) **1** one of several buttons on a panel for operating a computer.
2 a field in a record that is used to identify that record uniquely.
▸**v.** (**keys**, **keyed** /kēd/) enter or operate on (data) by means of a computer keyboard: [trans.] *she* ***keyed in*** *a series of commands* | [intrans.] *a hacker caused considerable disruption after* ***keying into*** *a vital database.*
DERIVATIVES **keyed adj.**; **key•less adj.**
ORIGIN Old English *cǣg*, *cǣge*, of unknown origin.

key•board /ˈkēˌbôrd/ ▸**n.** a panel of keys that operate a computer or typewriter.
▸**v.** [trans.] enter (data) by means of a keyboard.
DERIVATIVES **key•board•er n.**

key•chain drive /ˈkēCHān ˌdrīv/ ▸**n.** another term for **USB FLASH DRIVE.**

key•pad /ˈkēˌpæd/ ▸**n.** a miniature keyboard or set of buttons: *a numeric keypad.*

key•pal /ˈkēˌpæl/ ▸**n.** a person with whom one becomes friendly by exchanging e-mails; an e-mail pen pal: *one of the things on the Internet is e-mail. I mainly use it for sending e-mails to my keypals.*

ORIGIN 1990s: from *key* + *pal*, by analogy with *pen pal.*

key•punch /ˈkēˌpənCH/ ▸**n.** a device for transferring data by means of punching holes or notches on a series of cards or paper tape.

▸**v.** [trans.] put into the form of punched cards or paper tape by means of such a device.

DERIVATIVES **key•punch•er n.**

key•stroke /ˈkēˌstrōk/ ▸**n.** a single depression of a key on a keyboard, esp. as a measure of work: *program in simple text messages and send them with a single keystroke.*

key•word /ˈkēˌwərd/ ▸**n.** an informative word used in an information retrieval system to indicate the content of a document.

kill /kil/ ▸**v.** [trans.] stop (a computer program or process).

■ informal delete (a line, paragraph, or file) from a document or computer.

ORIGIN Middle English: probably of Germanic origin.

kill•er app /ˈkilər ˈæp/ ▸**n.** informal a feature, function, or application of a new technology or product that is presented as virtually indispensable or much superior to rival products: *e-mail is clearly the "killer app" that entices people online.*

kilo- ▸**comb. form** (used commonly in units of measurement) denoting a factor of 1,000.

ORIGIN via French from Greek *khilioi* 'thousand.'

kil•o•bit /ˈkiləˌbit/ ▸**n.** a unit of computer memory or data equal to 1,024 (2^{10}) bits.

kil•o•byte /ˈkiləˌbīt/ (abbr.: **Kb** or **KB**) ▸**n.** a unit of memory or data equal to 1,024 (2^{10}) bytes.

kludge /klo͞oj/ (also **kluge**) informal ▸**n.** a machine, system, or program that has been badly put together.

ORIGIN 1960s (originally U.S.): invented word, perhaps symbolic.

know•bot /ˈnōˌbät/ ▸**n.** a program on a network (esp. the Internet) that operates independently and has reasoning and decision-making capabilities. Also see **BOT**.

ORIGIN late 20th cent.: from *knowledgeable robot.*

knowl•edge /ˈnälij/ ▸**adj.** relating to organized information stored electronically: *the need for accurate self-assessment is magnified in the knowledge economy.*

knowl•edge base /ˈnälij ˌbās/ ▸ **n.** the underlying set of facts, assumptions, and rules that a computer system has available to solve a problem: *the knowledge base is a collection of facts and rules which relate to the subject of the expert system.*

KWIC /kwik/ ▸ **n.** [as adjective] keyword in context, denoting a database search in which the keyword is shown highlighted in the middle of the display, with the text forming its context on either side.
ORIGIN 1950s: abbreviation.

DEALING WITH SPAM

If you've used e-mail for any time at all, you've no doubt had your inbox deluged with messages pitching aphrodisiacs, mortgages, junk stocks, pornography, and substances that claim to help you lose weight in your sleep. But don't despair. Such messages don't have to be part of the Internet experience. You can avoid them, or certainly reduce the annoyance level, with a few simple measures.

Protect Your E-mail Address

Some spammers harvest e-mail addresses from the Internet by using a "spider" (also known as a robot or crawler)—a computer program that creeps through the World Wide Web collecting information you'd prefer to keep private. So the better you are at hiding your address from faceless prowlers, the less spam you will receive.

1. If your Internet Service Provider gives you the choice, create a long, fairly complicated address preceding the @; don't just use your first name: *donquixote@example.com* is better than *don@example.com.* Better still is *donquixotedelamancha@example.com.* It may be unwieldy, but your friends can always use a nickname—or just click on your address. Happily, longer addresses confuse spammers, who—without having exact user accounts in hand—use a "dictionary attack" to find them, testing the validity of every possible address from lists of common words and names: *arline@example.com, barbara@example.com,* and so forth.

2. Again, unless some online service makes it mandatory, don't use your e-mail address as part of a login name or identity. However convenient that might make it when logging in to eBay, it's equally convenient for spammers.

3. Use a second, public address for nonpersonal mail and for public posting. Create a free account with Hotmail, Yahoo, or Excite, or set up another screen name if you use either AOL or another Internet Service Provider that allows you to have multiple addresses. This keeps your main address private and exposes only your public address to spammers. Check the public account at least occasionally, in case something innocent has come in. Then massively delete the rest. It's important to learn how your e-mail client (the program you use to send and receive e-mail) handles thorough deleting; in Outlook Express, for example, you have to delete an e-mail you don't want from your inbox, delete it from the "Deleted" folder, and then compact folders before you are genuinely rid of it.

4. If you have a Web site, however much you want readers, customers, etc., to be able to communicate with you, don't post your unencoded e-mail address there. To block spammers from harvesting it, you can encode it in either of two ways—one technical and one surprisingly simple:

(1) Use a program designed to encode your address in hexadecimal or Javascript. For example, software advertising itself as "a simple anti-spam email link obfuscator in JavaScript" can be found at http://www.zapyon.de/spam-me-not/index.html(.) Another program, which offers you a choice of more than one encoding method, is at http://www.robertgraham.com/tools/ mailtoencoder. html(.) Using these programs, you can display normal text on your site that says nothing more revealing than "E-mail me" or "Click here."

(2) You can write your address so that only a human being would know how to replace and adjust certain parts: *dquixoteREPLACEWITHATexample.com,* or even just *dquixote at example dot com.*

5. Don't use your private e-mail address in the body or signature of posts to online forums, discussion boards, blogs, or newsgroups. If you must post an address, use your public one. Of course, you still want to take some care, or else it, too, will become overrun with spam.

6. Unless you're desperate for merchandise, choose not to receive *any* mail when registering for services online. If you can bring yourself to do it, turn off all newsletters, announcements, offers, deals, notices, and savings. You probably won't miss anything important, and you'll reduce the risk that someone dishonest will sell your e-mail address.

7. Never reply to spam or fall for its classic "click here to unsubscribe" instructions. This is important. That click will just add your address to new spam lists; most of those unsubscribe instructions are bogus—inserted to trap you into confirming with a single move of the mouse that your address really exists. In fact, you should be downright paranoid about clicking on *any* unsolicited e-mail (see "How to Protect Yourself from Hoaxes, Frauds, and Identity Theft," p. 130).

Filter your e-mail

No matter how careful you are, you're bound to receive spam. Well-meaning friends will send you and a dozen friends e-mail, leaving the addresses visible in the cc: field. That message will be forwarded and reforwarded, your e-mail address visible all the while, and eventually, someone unscrupulous will snag it and add it to their spam list.

Therefore, it's wise to set up your e-mail client to sort incoming messages automatically. Not only can you presort unread messages into organized folders, you can keep junk mail out of your inbox. Filtering is not perfect, however; good messages can be missorted, so make sure all suspect messages are merely moved, not deleted.

There are three common kinds of filtering.

1. **Algorithm filtering.** The most effective kind of algorithm filtering today is Bayesian; you actually train the filter to recognize spam by showing it both good messages and junk messages, telling it which is which so it learns to distinguish between them. It then starts to use this knowledge to judge incoming messages. If the filter makes a mistake, you tell the program, and the filter will take that error into account. Reports from users are good. Training takes time, but little by little the program learns to filter out some 99 percent of spam! Bayesian filtering is built into some mail programs and anti-spam applications.

2. **Keyword filtering.** Most online e-mail services, such as Hotmail, Yahoo, and Excite, offer keyword filtering, which looks for certain words that are likely to appear in spam messages: *Viagra, mortgage, loan, pharmacy, girls, deal, offer.* The more keywords that are matched, the greater the chance the message is spam. Stand-alone e-mail clients allow you to make rules yourself, which also can be used for keyword filtering, but trying to anticipate every spamlike keyword on your own is rather like trying to drain the ocean with a teaspoon.

3. **Whitelist filtering.** This is a harsh method when used alone, but when used in conjunction with the other two methods, it can work quite well. To set it up in any standalone e-mail program, just write a rule that says, "Any message received from someone not in my address book should be moved to the junk mail folder." As you can imagine, this method creates many false positives. It requires you keep your address book up to date, and it won't work well if you need to receive a lot of mail from strangers. It also requires that you wade through the junk mail folder regularly to make sure no valid messages were mistakenly filtered, which means you still have to look at all that spam you were trying to avoid in the first place!

Help! I'm Still Drowning in Spam

When even these precautions don't stop the flow of spam:

1. *Start over with a new e-mail address.* This may be the time to move to broadband. Consider getting an account with a service provider who offers server-side spam filtering, such as that done with SpamAssassin. Be very careful with your new address!

2. *Change to an e-mail client that has built-in spam filtering.* Microsoft Outlook, Microsoft Entourage, Apple Mail and the AOL connecting software have built-in junk mail filtering. Many of these will catch 75 percent of incoming spam, or more, right out of the box.

3. *Use a separate antispam application.* For Windows, SurfSecret SpamDrop is free for Outlook 2000, 2002 and XP, and support for Eudora and Outlook Express is expected. Both Norton and McAfee offer commercial antispam programs. SpamSieve is a great option for Apple Macintosh OS X and works with many different e-mail programs. Search on Download.com or versiontracker.com for other antispam options.

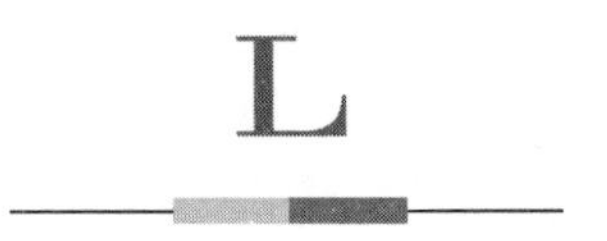

LAN /læn/ ▸ **abbr.** local area network.

lan•guage /ˈlæNGgwij/ ▸ **n.** a system of symbols and rules for writing programs or algorithms: *a new programming language.*
ORIGIN Middle English: from Old French *langage*, based on Latin *lingua* ‘tongue.’

lan•guage en•gi•neer•ing /ˈlæNGgwij ˌenjəˌni(ə)riNG/ ▸ **n.** any of a variety of computing procedures that use tools such as machine-readable dictionaries and sentence parsers in order to process natural languages for industrial applications such as speech recognition and speech synthesis.

lap•top /ˈlæpˌtäp/ (also **lap•top com•put•er** /ˈlæpˌtäp kəmˌpyo͞otər/) ▸ **n.** a notebook computer.

la•ser point•er /ˈlāzər ˌpointər/ ▸ **n.** a pen-shaped pointing device that contains a small diode laser that emits an intense beam of light, used to direct attention during presentations.

la•ser print•er /ˈlāzər ˌprintər/ ▸ **n.** a printer in which a laser forms a pattern of electrostatically charged dots on a light-sensitive drum, which attract toner (or dry ink powder). The toner is transferred to a piece of paper to create the printed document and fixed by a heating process.

last mile /ˈlæst ˈmīl/ ▸ **n.** the connection between a home or office and a telecommunications network, carrier, or delivery system: [as adjective] *last-mile wireless technology.*

LCD ▸ **abbr.** liquid crystal display.

least sig•nif•i•cant bit /ˈlēst sigˈnifikənt ˈbit/ (abbr.: **LSB**) ▸ **n.** the bit in a binary number that is of the lowest numerical value.

LED ▸ **abbr.** light-emitting diode, a semiconductor diode that glows when a voltage is applied.

leg•a•cy /ˈlegəsē/ ▸ **adj.** denoting software or hardware that has been superseded but continues to be included in new products due to consumer expectation or continued use: *there's an 80GB hard disk and legacy devices, such as the floppy drive, are available too.*

let•ter-qual•i•ty /ˈlet̲ər ˌkwälit̲ē/ ▸ **adj.** (of a printer attached to a computer) producing print of a quality suitable for business letters.

li•brar•y /ˈlīˌbrerē/ ▸ **n.** (pl. **li•brar•ies**) (also **soft•ware li•brar•y**) a collection of programs and software packages made generally available, often loaded and stored on disk for immediate use.

■ also (**pro•gram li•brar•y**) a stored collection of programming routines which can be called by a program as needed. Also see **DLL**.

ORIGIN late Middle English: via Old French from Latin *libraria* 'bookshop,' feminine (used as a noun) of *librarius* 'relating to books,' from *liber, libr-* 'book.'

LIFO /ˈlīfō/ ▸ **abbr.** last in, first out (with reference to methods of data storage). Compare with **FIFO**.

light gun /ˈlīt ˌgən/ ▸ **n.** a hand-held gunlike photosensitive device used in computer games, held to the display screen for passing information to the computer.

light pen /ˈlīt ˌpen/ ▸ **n. 1** a hand-held, penlike photosensitive device held to the display screen of a computer terminal for passing information to the computer.

2 a hand-held, light-emitting device used for reading bar codes.

line feed /ˈlīn ˌfēd/ ▸ **n.** a printer's advance of paper by one line when instructed by the document code sent by the computer.

■ the analogous movement of text on a computer screen.

line print•er /ˈlīn ˌprintər/ ▸ **n.** a machine that prints output from a computer a line at a time rather than character by character.

link /liNGk/ ▸ **n.** a code or instruction that connects one part of a program or an element in a list to another.

■ a hyperlink: *click on the link.* ■ contact by means of cable, telephone line, or radio waves: *a live audio webcast will be available via an Internet link.*

▸ **v.** make, form, or suggest a wireless or Internet connection: [trans.] *the newspaper has a policy against* ***linking to*** *its site.*

■ connect or join together: [trans.] *a Wi-Fi network wirelessly linking computers and other devices.*

ORIGIN late Middle English (denoting a loop; also as a verb in the sense 'connect physically'): from Old Norse *hlekkr*, of Germanic origin; related to German *Gelenk* 'joint.'

linked list /ˈliNGkt ˈlist/ ▸ **n.** an ordered set of data elements, each containing a link to its successor (and sometimes its predecessor).

link•er /ˈliNGkər/ ▸ **n.** a program used with a compiler or assembler to provide links to the libraries needed for an executable program.

Lin•ux /ˈlinəks/ ▸ **trademark** an operating system modeled on Unix, whose source code is publicly available at no charge.

ORIGIN 1990s: from the name of *Linus* Benedict Torvalds (b. 1969), a Finnish software engineer who wrote the first version of the system, + *-x*, as in *Unix*.

liq•uid crys•tal dis•play /ˈlikwid ˈkristl diˈsplā/ (abbr.: **LCD**) ▸ **n.** a form of visual display used for flat-panel displays, portable computers, and other devices, in which a layer of a liquid crystal is sandwiched between two transparent electrodes. Electric current alters the alignment of the liquid crystal molecules, affecting reflectivity or transmission of polarized light, creating the viewable images.

Lisp /lisp/ (also **LISP**) ▸ **n.** a high-level computer programming language devised for list processing.

ORIGIN 1950s: from *lis(t) p(rocessor).*

list•box /ˈlistˌbäks/ ▸ **n.** a box on the screen that contains a list of options, only one of which can be selected.

list proc•ess•ing /ˈlist ˈpräsesiNG/ ▸ **n.** the manipulation of data organized as lists.

LISTSERV /ˈlistˌsərv/ ▸ **trademark** a program to manage an electronic mailing list, often of people who form an interest group or wish to receive information of a specific type.

■ (also **list•serv**) the e-mail list managed by this program: *a recent post on the Tennessee Birds listserv.* ■ (also **list•serv**) any similar e-mail list application.

lit•tle en•di•an /'litl 'endēən/ ▸ see **ENDIAN**.

load /lōd/ ▸ **v.** insert (something) into a device so that it will operate: [trans.] *load the CD into the drive.*

■ insert something into (a device) so that it can be operated. ■ transfer (a program or data) to a hard drive: [trans.] *first you load the software onto your PC.* ■ transfer (a program or data) into memory, or into the central processor. ■ transfer programs or data into (a computer memory or processor): [intrans.] *it took 20 seconds for the home page to load.*

ORIGIN Old English *lād* 'way, journey, conveyance,' of Germanic origin; related to German *Leite*.

lo•cal /'lōkəl/ ▸ **adj.** denoting a device that can be accessed without the use of a network. Compare with **REMOTE**.

■ denoting a variable or other entity that is only available for use in one part of a program.

DERIVATIVES **lo•cal•ly adv.**

ORIGIN late Middle English: from late Latin *localis*, from Latin *locus* 'place.'

lo•cal ar•e•a net•work /'lōkəl 'e(ə)rēə 'net,wərk/ (abbr.: **LAN**) ▸ **n.** a computer network that links nearby devices, such as those in a room, a building, or adjacent buildings.

lo•cal bus /'lōkəl 'bəs/ ▸ **n.** a high-speed data connection directly linking peripheral devices to the processor and memory, allowing activities that require high data transmission rates such as video display.

lo•ca•tion /lō'kāsʜən/ ▸ **n.** a position or address in computer memory.

DERIVATIVES **lo•ca•tion•al adj.**

ORIGIN late 16th cent.: from Latin *locatio(n-)*, from the verb *locare*.

log /lôg/ ▸ **v.** (**logged**, **log•ging**) **1 log in** (or **on**) go through the procedures to begin use of a computer system, which includes establishing the identity of the user.

2 log off (or **out**) go through the procedures to conclude use of a computer system.

▸**n.** a record of errors or other events recorded by an active program.

ORIGIN Middle English, originally denoted a thin quadrant of wood loaded to float upright in the water, whence 'ship's journal' in which information from the log board was recorded.

log•ic /'läjik/ ▸ **n.** a system or set of principles underlying the arrangements of elements in a computer or electronic device so as to perform a specified task.

■ logical operations collectively.

ORIGIN late Middle English: via Old French *logique* and late Latin *logica* from Greek *logikē (tekhnē)* '(art) of reason,' from *logos* 'word, reason.'

log•i•cal op•er•a•tion /'läjikəl ˌäpə'rāsHən/ ▸ **n.** an operation that acts on binary numbers to produce a result according to the laws of Boolean logic (e.g., the AND, OR, and NOT functions).

log•ic bomb /'läjik ˌbäm/ ▸ **n.** a set of instructions secretly incorporated into a program so that if a particular condition is satisfied they will be carried out, usually with harmful effects.

log•ic cir•cuit /'läjik ˌsərkit/ ▸ **n.** a circuit for performing logical operations on input signals.

log•in /'lôgˌin/ (also **log•on**) ▸ **n.** an act of logging in to a computer system.

■ the information required to log in: *create a login.*

log•off /'lôgˌôf/ ▸ **n.** another term for **LOGOUT**.

log•on /'lôgˌän/ ▸ **n.** another term for **LOGIN**.

log•out /'lôgˌowt/ (also **log•off**) ▸ **n.** an act of logging out of a computer system.

look•up /'lo͝okˌəp/ ▸ **n.** [usually as modifier] the action of or a facility for systematic electronic information retrieval: *you need an online dictionary with fast phonetic lookup.*

loop /lo͞op/ ▸ **n.** a programmed sequence of instructions that is repeated until or while a particular condition is satisfied.

▸ **v.** [trans.] put into or execute a loop of computing instructions: *the program loops back on reaching a RETURN statement.*

loss•less /'lôslis/ ▸ **adj.** of or relating to data compression without loss of information: *a lossless format* | *lossless audio files.*

loss•y /'lôsē/ ▸ **adj.** of or relating to data compression in which unnecessary information is discarded: *the lossy method used in the image compression.*

love•ware /ˈləvˌwe(ə)r/ ▸ **noun** informal computer software that is distributed freely, with the developer asking for the users to think kindly of the developer or of a dedicatee in lieu of payment.

low-lev•el /ˈlō ˌlevəl/ ▸ **adj.** denoting a programming language or operations that are relatively close to machine code in form. Compare with **HIGH-LEVEL**.

lurk /lərk/ ▸ **v.** [intrans.] read communications on an electronic network without making one's presence known.

DERIVATIVES **lurk•er n.**

ORIGIN Middle English: perhaps from *lour* + the frequentative suffix *-k* (as in *talk*).

lurk•er /ˈlərkər/ ▸ **n.** a user of an Internet chat room or newsgroup who does not participate: *Mr. Rickard, of Boardwatch Magazine, estimates that there are five or six lurkers for each poster on a bulletin board.*

M

ma•chine code /mə'sʜēn ˌkōd/ (also **ma•chine lan•guage**) ▸ **n.** a computer programming language consisting of binary or hexadecimal instructions that a computer can respond to directly.

ma•chine-read•a•ble /mə'sʜēn ˌrēdəbəl/ ▸ **adj.** (of data or text) in a form that a computer can process: *machine-readable tickets.*

ma•chine trans•la•tion /mə'sʜēn transˌlāsʜən/ ▸ **n.** language translation carried out by a computer.

mac•ro /'mækrō/ ▸ **n.** (pl. **mac•ros**) (also **mac•ro in•struc•tion** /'mækrō inˌstrəksʜən/) a single instruction that expands automatically into a set of instructions to perform a particular task.

mac•ro vi•rus /'mækrō ˌvīrəs/ ▸ **n.** a macro containing a computer virus that attaches to a document and executes when the document is opened.

mail /māl/ ▸ **n.** e-mail.
▸ **v.** [trans.] send (someone) e-mail.
DERIVATIVES **mail•a•ble adj.**

mail bomb /'māl ˌbäm/ ▸ **n.** an overwhelmingly large quantity of e-mail messages sent to one e-mail address. Compare with **DENIAL-OF-SERVICE ATTACK.**
▸ **v.** (**mail-bomb**) [trans.] send an overwhelmingly large quantity of e-mail messages to (someone).

mail•box /'mālˌbäks/ ▸ **n.** a computer file in which e-mail messages received by a particular user are stored.

mail•er /'mālər/ ▸ **n.** a program that sends e-mail messages.

mail merge /'māl ˌmərj/ ▸ **n.** the automatic addition of names and addresses from a database to letters and envelopes in order to facilitate sending mail, esp. advertising, to many addresses.

main•frame /ˈmān,frām/ ▸ **n. 1** a large high-speed computer supporting numerous PCs or peripherals; a large server.
2 the central processing unit and primary memory of a computer.

mal•ware /ˈmal,we(ə)r/ ▸ **n.** software that is intended to damage or disable computers and computer systems: *protect your computer against viruses and other malware.*
ORIGIN blend of *malicious* and *software*.

man•age•ment in•for•ma•tion sys•tem /ˈmænijmənt infərˌmāsHən ˌsistəm/ ▸ **n.** (abbreviation **MIS**) a computerized information-processing system designed to support the activities of company or organizational management.

man•ag•er /ˈmænijər/ ▸ **n.** a program or system that controls or organizes a peripheral device or process: *a file manager.*

ma•nip•u•late /məˈnipyə,lāt/ ▸ **v.** [trans.] alter, edit, or move (text or data) on a computer.
DERIVATIVES **ma•nip•u•la•bil•i•ty n.**; **ma•nip•u•la•ble adj.**; **ma•nip•u•lat•a•ble** /-ˌlātəbəl/ **adj.**; **ma•nip•u•la•tion n.**
ORIGIN early 19th cent.: back-formation from earlier *manipulation*, from Latin *manipulus* 'handful.'

man page /ˈmæn ˌpāj/ ▸ **n.** a document forming part of the online documentation of a computer system: *make sure to check the man page.*
ORIGIN short for *manual page*.

mark•up /ˈmärk,əp/ ▸ **n.** a set of tags assigned to elements of a text to indicate their structural or logical relation to the rest of the text: *HTML markup* | [as adjective] *a markup language.*

mas•sive•ly par•al•lel /ˈmæsivlē ˈpærə,lel/ ▸ **adj.** (of a computer) consisting of many individual processing units, and thus able to carry out simultaneous calculations on a substantial scale: *a massively parallel computer with 168 processors.*

match /mæCH/ ▸ **n.** a string that fulfills the specified conditions of a computer search.
DERIVATIVES **match•a•ble adj.**
ORIGIN Old English *gemæcca* 'mate, companion,' of West Germanic origin.

MB ▸ **abbr.** (also **Mb**) megabyte: *an 800MB hard disk.*

Mbps ▸ **abbr.** ■ millions of bits per second. ■ megabits per second. ■ (**MBps**) megabytes per second.

Mbyte /ˈemˌbīt/ ▸ **abbr.** megabyte(s).

meat•space /ˈmētˌspās/ ▸ **n.** informal the physical world, as opposed to cyberspace or a virtual environment: *I'd like to know a little more before we talk about a get-together in meatspace.*

me•di•a /ˈmēdēə/ ▸ plural form of **MEDIUM**.

me•di•a card /ˈmēdēə ˌkärd/ ▸ **n.** a small cardlike information storage device that holds data in flash memory.

me•di•um /ˈmēdēəm/ ▸ **n.** (pl. **me•di•a**) **1** (also **storage medium**) a data storage material. Computer media may be magnetic (such as a hard disk), optical (such as a CD), or solid state (such as flash memory): *another item that makes a nice gift is some blank media, either CDs or DVDs.*

2 digital audio and video files: *streaming media.*

meg /meg/ ▸ **n.** (pl. same or **megs**) short for **MEGABYTE**.

meg•a•bit /ˈmegəˌbit/ ▸ **n.** a unit of data size or (when expressed per second) network speed, equal to one million or (strictly) 1,048,576 bits.

meg•a•byte /ˈmegəˌbīt/ (abbr.: **Mb** or **MB**) ▸ **n.** a unit of information equal to 2^{20} bytes or, loosely, one million bytes.

meg•a•flop /ˈmegəˌfläp/ ▸ **n.** a unit of computing speed equal to one million floating-point operations per second.

ORIGIN 1970s: back-formation from *megaflops* (see **MEGA-**, **-FLOP**).

meg•a•pix•el /ˈmegəˌpiksəl/ ▸ **n.** one million pixels; used as a measure of the resolution in digital cameras: [in comb.] *a 3.2-megapixel camera*

mem•o•ry /ˈmem(ə)rē/ ▸ **n.** (pl. **mem•o•ries**) the part of a computer in which data or program instructions can be stored for retrieval.

■ capacity for storing information in this way: *the module provides 16Mb of memory.*

ORIGIN Middle English: from Old French *memorie*, from Latin *memoria*, from *memor* 'mindful, remembering.'

mem•o•ry leak /ˈmem(ə)rē ˌlēk/ ▸ **n.** a failure in a computer program to deallocate discarded memory, causing impaired performance or failure: *repeatedly deleting toolbars using scripting impairs the performance of the program because of a "memory leak."*

mem•o•ry map•ping /ˈmem(ə)rē ˌmæpiNG/ ▸ **n.** a technique in which a computer treats peripheral devices as if they were located in the main memory.

mem•o•ry mod•ule /ˈmem(ə)rē ˌmäjo͞ol/ (also **mem•o•ry board** /ˈmem(ə)rē ˌbôrd/) ▸ **n.** a detachable board containing memory chips, which can be connected to a computer.

men•u /ˈmenyo͞o/ ▸ **n.** (pl. **men•us**) a list of commands or options, esp. one displayed on screen.

ORIGIN mid 19th cent.: from French, 'detailed list' (noun use of *menu* 'small, detailed'), from Latin *minutus* 'very small.'

men•u bar /ˈmenyo͞o ˌbär/ ▸ **n.** a horizontal bar, typically located at the top of the screen below the title bar, containing drop-down menus.

men•u-driv•en /ˈmenyo͞o ˌdrivən/ ▸ **adj.** (of a program or computer) used by making selections from menus.

merge /mərj/ ▸ **v.** [trans.] **1** incorporate revisions to a document to supersede the original: *if you answer "no" your changes will not be merged.*

2 combine (data or files) to produce a single entity: *The files were merged using the Patient Identification Code. . .as the common variable.*

mesh /mesH/ ▸ **n.** a set of finite elements used to represent a geometric object for modeling or analysis.

■ a computer network in which each computer or processor is connected to a number of others.

▸ **v.** [trans.] represent a geometric object as a set of finite elements for computational analysis or modeling.

DERIVATIVES **meshed adj.**; **mesh•y adj.**

ORIGIN late Middle English: probably from an unrecorded Old English word related to (and perhaps reinforced in Middle English by) Middle Dutch *maesche*, of Germanic origin.

mes•sage /ˈmesij/ ▸ **n.** an electronic communication generated automatically by a computer program and displayed on a monitor: *an error message.*

■ an item of e-mail.

ORIGIN Middle English: from Old French, based on Latin *missus*, past participle of *mittere* 'send.'

mes•sage board /'mesij ˌbôrd/ ▸ **n.** an Internet site where people can post and read messages, usually on a specific topic or area of interest. Compare with **BULLETIN BOARD.**

mes•sage box /'mesij ˌbäks/ ▸ **n.** a small box that appears on a computer screen to inform the user of something, such as the occurrence of an error.

mes•sage switch•ing /'mesij ˌswiCHiNG/ ▸ **n.** a mode of data transmission in which a message is sent as a complete unit and routed via a number of intermediate nodes at which it is stored and then forwarded. Compare with **PACKET SWITCHING.**

mes•sag•ing /'mesijiNG/ ▸ **n.** the sending and processing of e-mail by computer.

met•a /'metə/ ▸ **n.** short for **META KEY.**

ORIGIN 1980s: from *meta* in the sense 'beyond'.

met•a•da•ta /'metəˌdatə; -ˌdātə/ ▸ **n.** a set of data that describes and gives information about other data: *finally, they can search so-called metadata—extra information stored with the picture, including captions, the name of the photographer, the date of a picture, and so on.*

met•a•file /'metəˌfīl/ ▸ **n.** a piece of graphical information stored in a format that can be exchanged between different systems or software.

met•a key /'metə ˌkē/ ▸ **n.** a function key on a keyboard that is activated by simultaneously holding down a control key.

mi•cro /'mīkrō/ ▸ **n.** (pl. **mi•cros**) **1** short for **MICROCOMPUTER.** **2** short for **MICROPROCESSOR.**

mi•cro•brows•er /'mīkrōˌbrowzər/ ▸ **n.** a small Internet browser for use with cellular phones and other handheld devices.

mi•cro•code /'mīkrəˌkōd/ ▸ **n.** a very low-level instruction set that is stored permanently in a computer or peripheral controller and controls the operation of the device.

mi•cro•com•pu•ter /'mīkrōkəmˌpyo͞otər/ ▸ **n.** a personal desktop computer.

mi•cro•e•lec•tro•me•chan•i•cal /ˌmīkrō-iˌlektrōmə'kænikəl/ ▸ **adj.**

denoting systems or compenents relating to microscopic electronic machines that are typically built on computer chips: *optical true-time delay devices with microelectromechanical mirror arrays.*

DERIVATIVES **mi•cro•e•lec•tro•me•chan•ics** n.

mi•cro•in•struc•tion /ˌmīkrō-in'strəksHən/ ▸ **n.** a single instruction in microcode.

mi•cro•ker•nel /'mīkrōˌkərnl/ ▸ **n.** a small modular part of an operating system kernel that implements its basic features.

mi•cro•proc•es•sor /'mīkrəˌpräsesər/ ▸ **n.** another term for **PROCESSOR.**

DERIVATIVES **mi•cro•proc•ess•ing** n.

mi•cro•pro•gram /'mīkrōˌprōgrəm; -græm/ ▸ **n.** a microinstruction program that controls the functions of a central processing unit or peripheral controller of a computer.

▸ **v.** [trans.] use microprogramming with (a computer); bring about by means of a microprogram: *by 1980 virtually all computers were microprogrammed.*

DERIVATIVES **mi•cro•pro•gram•ma•ble** /ˌmīkrō'prōgræməbəl/ adj.; **mi•cro•pro•gram•mer** n.

mi•cro•pro•gram•ming /'mīkrōˌprōgrəmiNG; -græm-/ ▸ **n.** the technique of making machine instructions generate sequences of microinstructions in accordance with a microprogram rather than initiate the desired operations directly.

mi•cro•site /'mīkrəˌsīt/ ▸ **n.** an auxiliary Web site with independent links and address that is accessed mainly from a larger site: *to find out more about the winners and losers and where key industry figures rate Sulston, visit our Agenda Setters microsite.*

mid•dle•ware /'midlˌwe(ə)r/ ▸ **n.** software that occupies a position in a hierarchy between the operating system and the applications, whose task is to ensure that software from a variety of sources will work together correctly.

mi•grate /'mīˌgrāt/ ▸ **v.** [intrans.] change or cause to change from using one system to another.

■ [trans.] transfer (programs or hardware) from one system to another: *the product is designed to help users of older versions migrate to Windows XP.*

DERIVATIVES **mi•gra•tion** /mī'grāSHən/ n.

ORIGIN early 17th cent.: from Latin *migrat-* 'moved, shifted,' from the verb *migrare*.

MIME /mīm; 'em 'ī 'em 'ē/ ▸ **n.** a standard for formatting files of different types, such as text, graphics, or audio, so they can be sent over the Internet and seen or played by a Web browser or e-mail application.

ORIGIN late 20th cent.: an acronym for *multipurpose Internet mail extensions*.

min•i /'minē/ ▸ **n.** (pl. **min•is**) short for **MINICOMPUTER**.

ORIGIN 1960s: abbreviation.

min•i•com•pu•ter /'minēkəm,pyo͞otər/ ▸ **n.** in the past, a computer smaller than a mainframe, with several terminals, used by businesses and now largely replaced by servers.

min•i•tow•er /'minē,tow-ər/ (or **min•i-tow•er**) ▸ **n.** a small vertical case for a computer, or a computer mounted in such a case: *the desk has a compartment for a minitower* | [as adjective] *the minitower case is sturdy.*

MIPS /mips/ ▸ **n.** a unit of computing speed equivalent to a million instructions per second.

ORIGIN 1970s: acronym.

mir•ror /'mirər/ ▸ **n.** (also **mir•ror site** /'mirər ,sīt/) a site on a network that stores some or all of the contents from another site:

▸ **v.** [trans.] keep a copy of some or all of the contents of (a network site) at another site, typically in order to improve accessibility: *you can mirror the directory on a local server.*

■ [usu. as noun] (**mir•ror•ing**) store copies of data on (two or more hard disks) as a method of protecting it.

DERIVATIVES **mir•rored adj.**

ORIGIN Middle English: from Old French *mirour*, based on Latin *mirare* 'look at.'

MIS ▸ **abbr.** management information system.

mis•con•fig•ure /,miskən'figyər/ ▸ **v.** [trans.] [often as adj.] (**misconfigured**) configure (a system or part of it) incorrectly: *misconfigured Windows systems.*

DERIVATIVES **mis•con•fig•u•ra•tion** /,miskən,figyə'rāsʜən/ **n.**

mis•key /mis'kē/ ▸ **v.** (**mis•keys, mis•keyed**) [trans.] key (a word or piece of data) into a computer or other machine incorrectly.

mis•sion-crit•i•cal /ˈmisHən ˌkritikəl/ ▸ **adj.** (of hardware or software) vital to the functioning of an organization.

MO ▸ **abbr.** (of a disk or disk drive) magneto-optical.

mo•bile com•put•ing /ˈmōbəl kəmˈpyo͞otiNG/ ▸ the use or operation of computers away from a desk, office, or home, using a battery-powered notebook or handheld computer, as while traveling. *handheld PC users want to get the most from their mobile computing.*

mod /mäd/ ▸ **n.** (also **case mod**) a modification of a computer case, usually to add functions or accessories not intended or provided by the original manufacturer.

■ a modification of a computer by replacing the case with an unexpected type of container, often with windows to allow the electronic parts inside to be seen.

▸**v.** [trans.] modify a computer in this way.

DERIVATIVES **mod•der** /ˈmädər/ **n.**

mode /mōd/ ▸ **n.** a way of operating or using a system: *some computers provide several so-called processor modes.*

mo•dem /ˈmōdəm/ ▸ **n.** a combined device for modulation and demodulation between the digital data of a computer and the analog signal of a telephone line, allowing exchange of data between two remote computers and access to the Internet.

▸**v.** [trans.] send (data) by modem.

ORIGIN mid 20th cent.: blend of *modulator* and *demodulator*.

mod•ule /ˈmäjo͞ol/ ▸ **n.** any of a number of distinct but interrelated units from which a program may be built up or into which a complex activity may be analyzed.

mon•i•tor /ˈmänitər/ ▸ **n.** a computer display screen, usually a CRT or an LCD, used to view images generated by a computer.

mon•o•chrome /ˈmänəˌkrōm/ ▸ **adj.** (of a monitor or screen) displaying images in black and white or in varying tones of only one color.

ORIGIN mid 17th cent.: based on Greek *monokhrōmatos* 'of a single color.'

MOO /ˈmo͞o; ˈem ˈō ˈō/ ▸ **abbr.** MUD object-oriented, a MUD structured using an object-oriented programming language.

most sig•nif•i•cant bit /ˈmōst sigˈnifikənt ˈbit/ (abbr.: **MSB**) ▸ **n.** the bit in a binary number that is of the greatest numerical value.

moth•er•board /ˈməT͟Hərˌbôrd/ ▸ **n.** a printed circuit board containing the principal components of a computer or other device, with connectors into which other circuit boards can be attached.

mount /mownt/ ▸ v. [trans.] make (a disk or disk drive) available for use.

DERIVATIVES **mount•a•ble adj.**

ORIGIN Middle English: from Old French *munter*, based on Latin *mons, mont-* 'mountain.'

mouse /mows/ ▸ n. (pl. **mice** /mīs/ or **mous•es**) a small hand-held device that is dragged across a flat surface to move the cursor on a computer screen, typically having buttons that are pressed to control computer functions.

▸ v. [intrans.] informal use a mouse to move a cursor on a computer screen: *mouse your way over to the window and click on it.*

DERIVATIVES **mouse•like adj.**

ORIGIN Old English *mūs*, (plural) *mȳs*, of Germanic origin; related to Dutch *muis* and German *Maus*, from an Indo-European root shared by Latin and Greek *mus.*

mouse pad /ˈmows ˌpæd/ (also **mouse•pad**) ▸ n. a small piece of rigid or slightly resilient material on which a computer mouse is moved.

mouse po•ta•to /ˈmows pəˌtāt̲ō/ ▸ n. informal a person who spends large amounts of time operating a computer.

ORIGIN 1990s: on the pattern of *couch potato.*

mouse•trap /ˈmousˌtrap/ ▸ v. [trans.] (often as **mouse•trap•ping**) to block (a user's) efforts to exit from a Web site, usually one to which he or she has been redirected: *mousetrapping is a tactic commonly used by pornographic Web sites.*

MP3 ▸ n. a standard for compressing audio files, one of several standards used to reduce the storage space needed on a drive for a music file.

■ an audio file compressed using MP3 and stored on a computer, CD, or audio player.

ORIGIN late 20th cent.: from *Motion Pictures Experts Group Audio Layer 3.*

MPEG /ˈemˌpeg/ ▸ n. an international standard for encoding and compressing video images.

ORIGIN 1990s: from *Motion Pictures Experts Group.*

MP3 ▸ **n.** a standard for compressing audio files, used especially as a way of downloading music from the Internet.
ORIGIN 1990s: from **MPEG** + *Audio Layer-3.*

MS-DOS /ˌem ˌes ˈdäs/ ▸ **trademark** Microsoft disk operating system.

MTBF ▸ **abbr.** mean time between failures, a measure of the reliability of a device or system.

MUD /məd/ ▸ **n.** a computer-based text or virtual reality environment in which users adopt a character and interact with each other as well as with characters controlled by the computer.
ORIGIN late 20th cent.: from *multiuser dungeon* or *multiuser dimension.*

mul•ti•ac•cess /ˌməltēˈækˌses/ ▸ **adj.** (of a computer system) allowing the simultaneous connection of a number of terminals.

mul•ti•cast /ˈməltiˌkæst; ˌməltiˈkæst/ ▸ **v.** (past and past part. **mul•ti•cast**) [trans.] send data or transmit a signal to multiple selected recipients simultaneously: *during the day we'll multicast and during prime time, from probably 7-11 pm, we'll do our high-definition feed.*
▸ **n.** a set of data or an audio/video signal that is multicasted.

mul•ti•func•tion drive /ˈməltiˈfəNGkSHən ˈdrīv/ ▸ **n.** an optical disk drive that can perform several functions, such as read and record CDs and read DVDs, or read and record both CDs and DVDs.Compare with **COMBO DRIVE**.

mul•ti•func•tion print•er /ˈməltiˈfəNGkSHən ˈprintər/ ▸ **n.** a computer printer device that is also a scanner and copier and sometimes a fax machine.

mul•ti•me•di•a /ˈməltiˈmēdēə/ ▸ **adj.** (of a computer) able to process a variety of forms of data, including text, graphics, video, audio, and animation: *multimedia capabilities.*
■ (of a file or presentation) consisting of both audio and video.

mul•ti•play•er /ˈməltiˌplā-ər/ ▸ **n.** a multimedia computer and home entertainment system that integrates a number of conventional and interactive audio and video functions with those of a personal computer.
▸ **adj.** denoting a computer game designed for or involving several players.

mul•ti•proc•ess•ing /ˌməltiˈpräsesiNG/ (also **mul•ti•pro•gram•ming**) ▸ **n.** the running of two or more programs or sequences of instructions simultaneously by a computer with more than one central processor. Compare with **MULTITASKING**.

mul•ti•proc•es•sor /ˌməltiˈpräsesər/ ▸ **n.** a computer with more than one central processor.

mul•ti•ses•sion /ˈməltiˌseSHən/ ▸ **adj.** denoting a format for recording onto a CD or DVD over two or more separate sessions.

mul•ti•slack•ing /ˌməltiˈslakiNG/ ▸ **noun** informal the practice of using a computer at work for tasks or activities that are not related to one's job: *most employers tolerate a certain amount of multislacking.* See also **CYBERSLACKER**.

DERIVATIVES **mul•ti•slack•er n.**

ORIGIN *multi-* + *slacking* 'working slowly or lazily,' on the pattern of *multitasking,* the simultaneous execution of multiple computer tasks by a single processor.

mul•ti•task•ing /ˌməltiˈtæskiNG/ ▸ **n.** the simultaneous execution of more than one program or task by a single computer processor. Compare with **MULTIPROCESSING**.

■ the use of a computer to carry out several tasks simultaneously.

DERIVATIVES **mul•ti•task v.**

mul•ti•thread•ing /ˈməltiˈTHrediNG/ ▸ **n.** a technique by which a single set of code can be used by several processors at different stages of execution.

DERIVATIVES **mul•ti•thread•ed adj.**

mul•ti•us•er /ˌməltēˈyo͞ozər/ ▸ **adj.** [attrib.] (of a computer system) able to be used by a number of people simultaneously.

■ denoting a computer game in which several players interact simultaneously using the Internet or other communications.

mul•ti•ven•dor /ˌməltiˈvendər/ ▸ **adj.** [attrib.] denoting or relating to computer hardware or software products or network services from more than one supplier.

HOW TO PROTECT YOURSELF FROM HOAXES, FRAUDS, AND IDENTITY THEFT

There are two basic categories of Internet lies: *frauds*, where the object is to get your money, and *hoaxes*, where the primary object is just to pull your chain, but there can be far-reaching consequences as well. Hoaxes give the perpetrators ego gratification as they watch their creation spread throughout the Internet. Frauds may give the perpetrators your life savings, and give you months—maybe years—of hassle as you try to repair your credit record and retrieve your very identity.

Hoaxes—Hoaxes are spread by e-mail and come in an endless variety of guises. There are, for example, fake virus warnings, chain letters promising riches if you follow their instructions (or threatening dire consequences if you don't); urban myths about women in peril, dogs in microwaves, and hypodermic needles on theater seats; letters that tug at your heart strings or appeal to your greedy side; Internet petitions (often based on false information); and letters claiming that Bill Gates wants to give you money. Yeah, right.

Even the most "innocent" hoaxes are harmful. At the very least, they take up your time, and they try to get you to forward them to other people as well. If you forward a letter to just 40 people, and each of them does the same, and so on, then after just four steps, more than two and a half million copies will have been sent out. That's a lot of wasted time and wasted bandwidth.

These letters can also contain dangerous misinformation and bad advice. One example is a common letter advising women not to stop when pulled over by the highway patrol, but instead to dial #77 on their cell phones to talk to the police—a wrong number in 48 of the 50 states! Perhaps the most common example is the virus hoax—typically a letter forwarded by someone you know warning you that if you find a certain file on your computer it means you are infected with a virus. The letter advises you to delete the file immediately and then pass the warning along to everyone in your address book because you have probably infected them as well. The catch is that the file you are told to remove is a normal part of your operating system. To avoid being victimized by a virus hoax, rely on a reputable antivirus company, not e-mail from a friend. If you keep your antivirus program up to date it will almost certainly catch any real virus that comes along; never remove anything else that someone claims is a virus without first looking it up on an antivirus site like the one maintained by Symantec (http://securityresponse.symantec.com/avcenter/vinfodb.html) or McAfee (http://vil.mcafee.com/). If something is a real virus, they will tell you how to remove it safely.

The common element in all hoaxes is that they try to get you to pass the letter on to others—making you one of the perpetrators of the hoax. Once launched, hoaxes depend on well-meaning dupes to keep them in circulation—people who react impulsively and who don't know or care enough to check on the accuracy of the information they forward.

To avoid being a dupe yourself, view all letters that are forwarded to you with skepticism. Common tip-offs that a letter is a hoax include use of ALL CAPS and MULTIPLE EXCLAMATION POINTS!!!, bad spelling and grammar, information supposedly heard from somebody who heard it from somebody (and with no verifiable details)—urging you to "forward this to everyone you care about," and even insistence that "this is not a hoax!" But virtually all chain letters *are* hoaxes, and forwarding them to people who haven't asked for them is simply spamming. If you haven't actually verified the information, just don't pass it on. As they say at www.purportal.com, "Search before you forward." That portal makes it easy—just enter a couple of key words from the letter and the site will link you to Web sites that sort out fact from fiction. A well organized and very readable general resource on hoaxes is the government-sponsored "Hoaxbusters" site at http://hoaxbusters.ciac.org/.

Frauds—Frauds are more complicated than "mere" hoaxes because they have to trick you into giving out money or confidential financial information. They may begin with an e-mail, but often involve an entire Web site as well. Enterprising criminals have lost no time adapting every possible swindle, con, and flim flam to the Internet age. These include investment scams, pyramid schemes, phony business opportunities, Nigerian money-moving frauds, and sales of counterfeit, ineffective, or even dangerous products.

Although these frauds were around long before we had PCs, the anonymity, speed, volume, and global reach of the Internet enable the perpetrators to evade law enforcement efforts while reaching vast numbers of victims. Be extremely leery of any financial proposition that plays on your emotions or tempts you with high or "guaranteed" profits. A great place to learn more about the many kinds of fraud to watch out for is the Federal Trade Commission's Web site at http://www.ftc.gov/bcp/menu-internet.htm.

The rise of the Internet and electronic commerce has not only facilitated old frauds, but also opened the door to new ones. For example, you might be enticed to download a "viewer" or "dialer" for access to "free" adult content, and later find hundreds or thousands of dollars added to your telephone bill for long distance calls made, without your knowledge, from your computer to remote islands that charge outrageous phone rates. Say "No!" to adult sites that want to install something on your computer.

But that scam is small potatoes compared with highly sophisticated e-mail-based schemes that can lead to credit card theft and, ultimately, *identity theft*—the use of your identity to engage in a whole range of financial transactions, leaving you with the bill. The most ingenious and insidious new scheme is *phishing*, in which you are lured to a counterfeit Web site by an e-mail that purports to be from your bank or some other institution that you trust (like AOL or PayPal). The letter may say that you need to update your account information in order to keep the account active, or that you must provide information so that a check that has been received for you can be deposited. If you click on the link provided in the e-mail, you are taken to a *spoofed site*—a Web site that looks like the company's real site but is actually controlled by criminals—containing a form calling for sensitive information like your bank account number, Social Security number, credit card number, password, or PIN. Aside from the occasional lapse in syntax or spelling, the request may seem credible, and the spoofed site may look stunningly like the real one, complete with genuine logos, familiar typefaces and images, and even a reassuring link to the privacy policy on the company's real site. You may even see the company's actual URL in your browser's address bar.

The best way to avoid becoming phish bait is to avoid clicking on such e-mails in the first place. If you think the e-mail may be legitimate, go to the site directly or call customer service for more information—using addresses and numbers that you know are genuine. However, if you do use the e-mail link, you should try to verify the site's security certificate. First see if there are *yellow security icons* in your browser's status bar (see How to Shop Safely Online, p. 143). While these usually signal that the page you are on is secure and that your personal information will be encrypted, such an icon may show up on a spoofed site, in which case it may have been purchased by the scam artists. Try double-clicking on it to bring up the name of the security certificate's owner. If nothing appears, you will want to check the site's legitimacy in some other way. In recent versions of Internet Explorer, from any page that claims to be secure, click on File > Properties and then on the "Certificates" button. When the certificate comes up, make sure that "Issued to" is followed by the name of the correct site. If you see any other name, leave the site without typing anything into it.

If you receive an e-mail that smells phishy, report it to the legitimate site owner (e.g., PayPal or your bank), and forward the e-mail to uce@ftc.gov. If you've already made the mistake of giving out personal information, file a complaint with the Federal Trade Commission by going to www.ftc.gov/ftc/consumer.htm and clicking on "FILE A COMPLAINT." Then find out what to do next at the FTC's Identity Theft Web site (www.ftc.gov/idtheft). But take heart: people are working on the problem. See more about phishing and what is being done to combat it at www.antiphishing.org/.

N

nag•ware /ˈnæg,we(ə)r/ ▸ **n.** informal computer software that is free for a trial period during which frequent reminders appear on screen asking the user to register and pay for the program in order to continue using it when the trial period is over: *my recent Weblog item about QuickBooks nagware elicited a number of comments from readers pointing out yet other shenanigans they've seen.*

nas•ty•gram /ˈnæstē,græm/ ▸ **n.** an offensive or threatening electronic communication: *late last year the company began sending out some nastygrams to subscribers they identified as being wireless broadband connection sharers.*

na•tive /ˈnātiv/ ▸ **adj.** designed for or built into a given system, esp. denoting the language associated with a given processor, computer, or compiler, and programs written in it.

DERIVATIVES **na•tive•ly** adv.; **na•tive•ness** n.

ORIGIN late Middle English: from Latin *nativus*, from *nat-* 'born,' from the verb *nasci*.

nav•i•gate /ˈnavi,gāt/ ▸ **v.** move around a Web site, file, the Internet, etc.: *we've added features that make our site much easier to navigate.*

ne•mat•ic /niˈmætik/ ▸ **adj.** relating to or denoting a state of a liquid crystal in which the molecules are oriented in parallel but not arranged in well-defined planes: *nematic liquids are found in many LCD displays, especially ones used in consumer electronics.*

▸ **n.** a nematic substance.

ORIGIN early 20th cent.: from Greek *nēma, nēmat-* 'thread' + **-IC**.

net /net/ ▸ **n.** (**the Net**) the Internet.

■ a network of interconnected computers: *a computer news net.*

ORIGIN Old English *net, nett,* of Germanic origin; related to Dutch *net* and German *Netz.*

net•i•quette /ˈnetikit/ ▸ **n.** the correct or acceptable way of communicating on the Internet.

ORIGIN 1990s: blend of **NET** and *etiquette.*

net•i•zen /ˈnetəzən/ ▸ **n.** a user of the Internet, esp. a habitual or avid one: *learn how to be a responsible netizen.*

ORIGIN 1990s: blend of **NET** and *citizen.*

net•work /ˈnetˌwərk/ ▸ **n.** a number of computers interconnected so they can share files and use of other devices, such as printers and scanners, that are also connected to the network: *a wireless home network.*

▸ **v.** [trans.] link (computers and other devices) to operate interactively: [as adjective] (**networked**) *a networked printer.*

net•work ap•pli•ance /ˈnetˌwərk əˌplīəns/ ▸ **n.** a relatively low-cost computer designed chiefly to provide Internet access and without the full capabilities of a standard personal computer.

net•work com•put•er /ˈnetˌwərk kəmˌpyo͞otər/ ▸ **n.** ■ a personal computer with reduced functionality intended to be used to access services on a network.

neu•ral com•pu•ter /ˈn(y)o͝orəl kəmˈpyo͞otər/ ▸ **n.** a computer that uses neural networks based on the human brain: *the Ricoh machine might be the first complete neural computer, as opposed to a more general-purpose computer containing neural network chips or software.*

DERIVATIVES **neu•ral com•put•ing n.**

neu•ral net•work /ˈn(y)o͝orəl ˈnetˌwərk/ (also **neu•ral net**) ▸ **n.** a computer system modeled on the human brain and nervous system.

neu•ro•com•pu•ter /ˈn(y)o͝orōkəmˌpyo͞otər/ ▸ **n.** another term for **NEURAL COMPUTER.**

new•bie /ˈn(y)o͞obē/ ▸ **n.** (pl. **new•bies**) an inexperienced newcomer, esp. in computing.

news•serv•er /n(y)o͞ozˌsərvər/ (also **new•serv•er** /ˈn(y)o͞o(z) ˌsərvər/) ▸ **n.** an Internet-connected server that receives and disseminates messages for a newsgroup.

news•feed /ˈn(y)o͞ozˌfēd/ ▸ **n.** an electronic transmission of news, as from a broadcaster or an Internet newsgroup: *full Internet capabilities are available, such as Usenet newsfeeds for more than 7,000 newsgroups.*

news•group /ˈn(y)o͞ozˌgro͞op/ ▸ **n.** an Internet-based forum devoted to discussing a particular topic: *then I discovered the verizon.adsl newsgroup and there are about a dozen serious users complaining about the problem all over Hampton Roads.*

■ the subscribers to such a group.

news•read•er /ˈn(y)o͞ozˌrēdər/ ▸ **n.** a computer program for reading e-mail messages posted to newsgroups.

node /nōd/ ▸ **n.** a piece of equipment, such as a PC or peripheral, attached to a network.

DERIVATIVES **nod•al** /ˈnōdl/ **adj.**

ORIGIN late Middle English (denoting a knotty swelling or a protuberance): from Latin *nodus* 'knot.'

noise•less /ˈnoizlis/ ▸ **adj.** accompanied by or introducing no random fluctuations that would obscure the real signal or data.

non•dig•i•tal /nänˈdijitl/ ▸ **adj.** **1** not represented by numbers, especially binary codes; not digitized: *nondigital items have only their location information (catalog records) in the digital library, as it happens in a traditional automated library situation.*

2 not using the Internet or computers: *nondigital submissions will be accepted only until February 1st.*

non•pro•pri•e•tar•y /ˌnänprəˈprīəˌterē/ ▸ **adj.** (esp. of computer hardware or software) conforming to standards that are in the public domain or are widely licensed, and so not restricted to one manufacturer.

non•res•i•dent /nänˈrezidənt/ ▸ **adj.** (of software) not kept permanently in memory but available to be loaded from a backing store or external device: *if you want to use a nonresident font, you can manually download it.*

DERIVATIVES **non•res•i•dence n.**

non•so•lid col•or /nänˈsälid ˈkələr/ ▸ **n.** a color simulated by a pattern of dots of other colors, extending the range of colors available.

non•vol•a•tile /nänˈvälət̲l/ ▸ **adj.** (of a computer's memory) retaining data even if there is a break in the power supply.

NOT /nät/ ▸ **n.** a Boolean operator with only one variable that is true when the variable is false and vice versa. See usage note at **BOOLEAN.**

note•book /ˈnōtˌbo͝ok/ (also **lap•top**) ▸ **n.** a small battery-powered portable computer with most of the functionality of a desktop computer.

note•pad /ˈnōtˌpæd/ ▸ **n.** (also **note•pad com•put•er**) another tern for a **HANDHELD.**

null /nəl/ ▸ **adj.** (of a set or matrix) having no elements, or only zeros as elements.

■ having or associated with the value zero.

ORIGIN late Middle English: from French *nul*, *nulle*, from Latin *nullus* 'none,' from *ne* 'not' + *ullus* 'any.'

null char•ac•ter /ˈnəl ˌkæriktər/ ▸ **n.** a character denoting nothing, usually represented by a binary zero.

null link /ˈnəl ˈliNGk/ ▸ **n.** a reference incorporated into the last item in a list to indicate that there are no further items in the list.

num•ber crunch•er /ˈnəmbər ˌkrənCHər/ (also **num•ber-crunch•er**) ▸ **n.** informal a computer or software capable of performing rapid calculations with large amounts of data.

DERIVATIVES **num•ber crunch•ing n.**

O

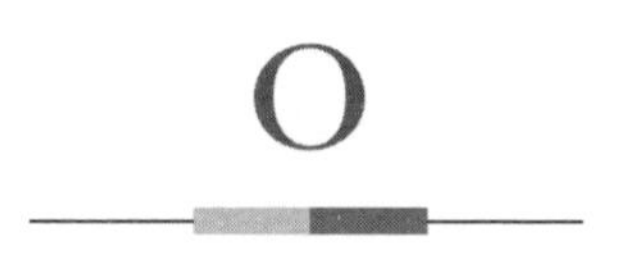

ob•ject /ˈäbjikt/ ▸ **n.** a data construct that provides a description of something that may be used by a computer (such as a processor, a peripheral, a document, or a data set) and defines its status, its method of operation, and how it interacts with other objects.
ORIGIN late Middle English: from medieval Latin *objectum* 'thing presented to the mind,' neuter past participle (used as a noun) of Latin *obicere*, from *ob-* 'in the way of' + *jacere* 'to throw'; the verb may also partly represent the Latin frequentative *objectare*.

ob•ject code /ˈäbjikt ˌkōd/ ▸ **n.** code produced by a compiler or assembler.

ob•ject lan•guage /ˈäbjikt ˌlæNGgwidj/ ▸ **n.** a language into which a program is translated by means of a compiler or assembler.

ob•ject-o•ri•ent•ed /ˈäbjikt ˌôrēˌentid/ ▸ **adj.** (of a programming language) using a methodology that enables a system to be modeled as a set of objects that can be controlled and manipulated in a modular manner. *object-oriented programming* | *software experts and object-oriented technology gurus.*
DERIVATIVES **ob•ject o•ri•en•ta•tion** /ˈäbjikt ˌôrēenˈtāSHən/ **n.**

OCR ▸ **abbr.** ■ optical character recognition. ■ optical character reader.

OEM ▸ **abbr.** original equipment manufacturer, an organization that makes devices from component parts bought from other organizations.

off•line /ˈôfˈlīn/ (also **off-line**) ▸ **adj.** not connected to, using, or involving the Internet: *offline research.*

■ not controlled by or directly connected to a computer or external network:

▸ **adv.** (also **off line**) while not using the Internet: *e-mail is not available when you are offline.*

■ (**go offline**) finish using or being connected to the Internet. ■ while not directly controlled by or connected to a computer or external network.

off•load /ˈôfˌlōd/ (also **off-load**) ▸ **v.** [trans.] move (data or a task) from one processor to another in order to free the first processor for other tasks: *a system designed to offload the text on to a host computer.*

oh•no•sec•ond /ˈōˈnōˌsekənd/ ▸ **n.** informal a moment in which one realizes that one has made an error, typically by pressing the wrong key: *you may have heard of the "ohnosecond" when you realise that the shit is about to hit the fan.*

OLE /ōˈlā; ˈō ˈel ˈē/ ▸ **abbr.** object linking and embedding, denoting a set of techniques for transferring an object from one application to another.

on-board /ˈän ˌbôrd/ ▸ **adj.** [attrib.] denoting or controlled from a facility or feature incorporated into the main circuit board of a computer or computerized device.

one-time pad /ˈwən ˌtīm ˈpæd/ ▸ **n.** an encryption technique based on the one-time use of a randomly generated key unique to each encryption and requiring the matching key to make decryption possible.

on•line /ˈänˌlīn/ (also **on-line**) ▸ **adj.** connected to, using, or involving the Internet: *the ease and convenience of online shopping.*

■ controlled by or connected to another computer or to a network.

▸ **adv.** (also **on line**) while connected to the Internet: *applicants must have experience working online.*

■ (**go online**) start using or being connected to the Internet. ■ while connected to or under computer control.

on-screen /ˈän ˈskrēn/ (also **on screen** or **on•screen**) ▸ **adj.** & **adv.** making use of or performed with the aid of a computer display screen: [as adjective] *on-screen editing.*

on the fly /ˈän T͟Hə ˈflī/ ▸ **n.** during the running of a computer program without interrupting it.

OOP /o͞op; 'ō 'ō 'pē/ ▸ **abbr.** object-oriented programming.

o•pen /'ōpən/ ▸ **v.** [trans.] take the action required to make ready for use: *click twice to open a file.*

ORIGIN Old English *open* (adjective), *openian* (verb), of Germanic origin; related to Dutch *open* and German *offen*, from the root of the adverb *up*.

o•pen sys•tem /'ōpən 'sistəm/ ▸ **n.** a computer system in which the components and protocols conform to standards independent of a particular supplier.

op•er•at•ing sys•tem /'äpəˌrātiNG ˌsistəm/ ▸ **n.** the software that supports a computer's basic functions, such as scheduling tasks, executing applications, and controlling peripherals.

op•er•a•tion /ˌäpə'rāSHən/ ▸ **n.** a process in which a number, quantity, expression, etc., is altered or manipulated according to formal rules, such as those of addition, multiplication, and differentiation.

ORIGIN late Middle English: via Old French from Latin *operation(n-)*, from the verb *operari* 'expend labor on.'

op•er•a•tor /'äpəˌrātər/ ▸ **n.** a symbol or function denoting an operation (e.g., ×, +).

op•ti•cal char•ac•ter rec•og•ni•tion /'äpətikəl 'kæriktər rekəg ˌniSHən / (abbr.: **OCR**) ▸ **n.** the identification of printed characters using a scanner and computer software, thus turning printed documents into electronically editable text.

op•ti•cal disk /'äptikəl 'disk/ ▸ **n.** a disk on which data bits are represented by tiny reflective and nonreflective areas that can be read by a sensor as reflected laser light.

■ a compact disk. ■ a DVD.

op•ti•cal mouse /'ätikəl 'mows/ ▸ **n.** a mouse using the reflected light of an LED hitting a sensor to track changes in its position.

OR /ôr/ ▸ **n.** a Boolean operator that evaluates as true if at least one argument is true, and otherwise has a value of false. See usage note at **BOOLEAN**.

OS ▸ **abbr.** operating system.

out•put /'owtˌpo͝ot/ ▸ **n.** data transmitted from a computer to another computer or a peripheral device.

■ the product produced by a peripheral device: *composite video output | the quality of the output from the printer is very good.*

▸ v. (**out•put•ting**; past and past part. **out•put** or **out•put•ted**) [trans.] produce, deliver, or supply (data) using a computer or other device: *you can output the image directly to a video recording system.*

o•ver•lay /ˈōvərˌlā/ ▸ **n. 1** a graphical computer image that can be superimposed on another.

2 the process of transferring a block of program code or other data into internal memory, replacing what is already stored.

■ a block of code or other data transferred in such a way.

o•ver•type /ˌōvərˈtīp/ ▸ v. [trans.] type over (another character): *overtype it with the correct number and press Enter.*

▸ n. a facility or operating mode allowing overtyping.

o•ver•write /ˌōvərˈrīt/ ▸ v. (past **o•ver•wrote** /-ˈrōt/; past part. **o•ver•writ•ten** /-ˈritn/) [trans.] destroy (data) or the data in (a file) by entering new data in its place: *an entry stating who is allowed to overwrite the file.*

■ another term for **OVERTYPE**.

HOW TO SHOP SAFELY ONLINE

Whether you're a compulsive gadget freak, an eBay addict, or a restrained occasional buyer, the Internet is a shopper's paradise—open 24/7, offering unlimited choices, and giving us the luxury of shopping—through sleet and snow or summer heat—from our own homes. But at no time have caveats been more important for emptors than now. Devious and ingenious scam artists are just waiting to pounce on anyone who doesn't take appropriate precautions. A few sound practices can help assure that you get what you pay for, get it on time, and don't lose your shirt in the bargain.

1. **Use a secure browser**—Believe it or not, you can't take browsers for granted. Be sure that you have the latest version of yours, including the latest updates and security patches, and that you've set your browser to notify you when you are entering or leaving a secure site. Be sure as well that it complies, as major browsers do (e.g., Netscape and Internet Explorer), with common industry security standards, like SSL and SET technology (see 2 just below).

2. **Check out your vendor**—Try to use vendors you're familiar with. Many well-known bricks-and-mortar stores are now "clicks-and-mortar" operations, with both physical stores and sites on the Internet. If you must buy from a place you don't know, check the site's security. A secure site will use VeriSign's

Secure Sockets Layer technology (SSL), displaying a locked padlock at the bottom of the screen, Secure Electronic Transaction technology (SET), which shows an unbroken key, or a VeriSign logo indicating that you are at a secure Site. These are not just decorative icons. They are intended to ensure that all personal information you submit to the site will be scrambled en route through the ether and decrypted only when it reaches the licensed merchant. However, this is just the first step. Particularly if you clicked on an ad or a URL sent to you in an e-mail, and therefore did not type the URL in the address bar yourself, you should verify the security certificate (see How to Protect yourself from Frauds, Scams, and Identity Theft, p. 130). You can also check vendors and manufacturers with the Better Business Bureau (www.bbb.org) or with one of the many sites specializing in user ratings for stores and products, such as: (1) http://bizrate.com/ratings_guide/guide.xpml, (2) http://www.epinions.com/, (3) http://www.planetfeedback.com/consumer/, and (4) http://www.complaints.com. You may get so caught up in reading complaints that you'll forget about shopping; think of the money you'll save!

3. **Be inventive with your passwords**—When you establish a new account, create a "strong" password that can't easily be hacked. Common words and names, whether spelled forwards or backwards, can be unearthed in an instant by unscrupulous thieves using "dictionary" software, so your child's name or the name of your pet are out. Your birthday or telephone number should also be avoided, and don't even think about using your Social Security number. Ideally, you should use at least six characters—combining upper- and lowercase letters, numbers from 1 to 9, and—I swear—symbols like @#$%^&*. Unfortunately, some sites allow lowercase letters only, or just letters and numbers, in which case you'll have to be creative—perhaps devising an acronym or using a nonsense sequence you can remember, like the initials of several friends mixed in with their ages. Vary your passwords, change them frequently, never give them to any-

one else, and for heaven's sake don't put them on sticky notes on your office monitor.

4. **Read the documentation**—If you're new to a site, read whatever you can find there about shipping rates (and while you're at it, look for chances to get free shipping). Check return policies and restocking fees, and—above all—the company's privacy policy as regards your personal information. You'll want to know, for example, if they sell their mailing lists. When establishing an account, it's wise to leave blank any information, such as your home phone number, that isn't marked on signup forms as *required.* Read to the bottom of every page and check for check marks or pre-filled-in buttons. You may not want to accept a default that gives the seller permission to pass your e-mail address along to third party "partners."

5. **Keep records**—At the very least, (1) print out a hard copy of your order confirmation and (2) save it electronically in a Shopping folder on your hard drive. Better, print out every phase of the order—from product selection to checkout. Keep this kind of information in accessible file folders so that you can handle possible returns, a vendor's failure to deliver satisfactory items, and a pressing need in March and April for information relevant to tax returns.

6. **Auction with care**—According to the Federal Trade Commission, auction fraud is the number-one complaint received from Internet shoppers; a variety of scams—perpetrated by buyers and sellers alike—can attack you on eBay and other online auction sites. If you're a buyer, you can pay for an item that never arrives or that is defective. If you're a seller, you can send a buyer an item after receiving a bad check. And yet, tons of safe and satisfactory transactions are conducted through these sites every day. EBay is addictive for a reason; buyers can find items there that are otherwise unobtainable and sellers can profitably dispose of unwanted items. The point

is: you have to do your research and proceed with caution. The eBay site has an excellent Security Center page (http://pages.ebay.com/securitycenter/) that, if read carefully (this means clicking on all of the sub-pages) and *followed*, will help make you a happy auctioneer or auctionee.

7. **Pay by credit card**—It is probably *more* safe to use your credit card online than in, say, a restaurant. In fact, credit cards offer the safest way to pay online. Your credit card transactions are protected under the Fair Credit Billing Act; in the unlikely event that someone else does acquire your number, you are liable for a maximum of $50. However, since roughly two percent of online consumers *have* complained that their credit card number was stolen and used fraudulently, new technology by some card issuers presents a welcome innovation. It allows card holders to create single-use or substitute numbers for individual transactions. As this technology spreads, a large sigh of relief will be heard echoing through cyberspace. One more bit of personal information that you don't have to divulge.

P

P2P /ˈpē tə ˈpē/ ▸ **abbr.** peer-to-peer.

pack•age /ˈpækij/ ▸ **n.** a collection of programs or subroutines with related functionality.

pack•e•tize /ˈpækiˌtīz/ ▸ **v.** [trans.] partition or separate (data) into units for transmission in a packet-switching network, such as the telephone system: *this layer packetizes and reassembles messages.*

pack•et net•work /ˈpækit ˌnetwərk/ ▸ **n.** a data transmission network using packet switching.

pack•et switch•ing /ˈpækit ˌswiCHiNG/ ▸ **n.** a mode of data transmission in which a message is broken into a number of parts that are sent independently, over whatever route is optimum for each packet, and reassembled at the destination. Compare with **MESSAGE SWITCHING.**

pack•ing den•si•ty /ˈpækiNG ˌdensit̲ē/ ▸ **n.** the density of stored information in terms of bits per unit occupied of its storage medium.

page /pāj/ ▸ **n.** a section of stored data, esp. that which can be displayed on a screen at one time.

▸ **v.** [intrans.] (**page through**) move through and display (text) one page at a time.

■ [usu. as noun] (**paging**) divide (a piece of software or data) into sections, keeping the most frequently accessed in main memory and storing the rest in virtual memory.

ORIGIN late 16th cent.: from French, from Latin *pagina*, from *pangere* 'fasten.'

paint /pānt/ ▸ **n.** the function or capability of producing graphics, esp. those that mimic the effect of real paint: [as adjective] *a paint program.*

▸ v. [trans.] create (a graphic or screen display) using a paint program.
ORIGIN Middle English: from *peint* 'painted,' past participle of Old French *peindre*, from Latin *pingere* 'to paint.'

pal•ette /'pælit/ ▸ n. (in computer graphics) the range of colors or shapes available to the user.
ORIGIN late 18th cent.: from French, diminutive of *pale* 'shovel,' from Latin *pala* 'spade.'

Palm Pi•lot /'päm 'pīlət/ ▸ trademark a hand-held computer.

palm•top /'päm,täp/ ▸ n. another term for HANDHELD.

pane /pān/ ▸ n. a separate defined area within a window for the display of, or interaction with, a part of that window's application or output.
ORIGIN late Middle English (originally denoting a section or piece of something, such as a fence or strip of cloth): from Old French *pan*, from Latin *pannus* 'piece of cloth.'

pa•per tape /'pāpər 'tāp/ ▸ n. paper in the form of a long narrow strip into which holes can be punched, used in computer systems in the past to enter data or instructions.

par•al•lel port /'pærəlel 'pôrt/ ▸ n. a connector for a device that sends or receives several bits of data simultaneously on multiple wires.

par•al•lel proc•ess•ing /'pærə,lel 'präsesiNG/ ▸ n. a mode of computer operation in which a process is split into parts that execute simultaneously on different processors attached to the same computer: *some industries require very large databases and massive parallel processing capabilities.*

pa•ram•e•ter /pə'ræmitər/ ▸ n. (in the execution of a program) a value assigned to a variable passed to the program by a user or another program: *a parameter that accepts the name of a U.S. state.*
ORIGIN mid 17th cent.: modern Latin, from Greek *para-* 'beside' + *metron* 'measure.'

par•i•ty bit /'pæritē ,bit/ ▸ n. a bit that acts as a check on a set of binary values, calculated in such a way that the number of 1s in the set plus the parity bit should always be even (or occasionally, should always be odd).

parse /pärs/ ▸ v. [trans.] analyze (a string or text) into logical syntactic components, typically in order to test conformability to a logical grammar.

▸**n.** an act of or the result obtained by parsing a string or a text.

ORIGIN mid 16th cent.: perhaps from Middle English *pars* 'parts of speech,' from Old French *pars* 'parts' (influenced by Latin *pars* 'part').

pars•er /'pärsər/ ▸ **n.** a program for parsing.

par•ti•tion /pär'tisHən/ ▸ **n.** each of a number of portions into which some operating systems divide memory or hard disk storage capacity.

▸**v.** [trans.] divide storage capacity on a hard disk into portions.

ORIGIN late Middle English: from Latin *partitio(n-)*, from *partiri* 'divide into parts.'

par•ti•tion•er /pär'tisHənər/ ▸ **n.** a piece of software for apportioning space on a hard disk. A **hard partitioner** does this prior to formatting (i.e., permanently), while a **soft partitioner** does it after formatting.

Pas•cal /pæ'skæl/ (also **PASCAL**) ▸ **n.** a high-level structured computer programming language used for teaching and general programming.

pass /pæs/ ▸ **n.** a single scan through a set of data or a program.

ORIGIN Middle English: from Old French *passer*, based on Latin *passus* 'pace.'

Pass•face /'pasˌfās/ ▸ **trademark** **1** a security system in which a user must recognize pictures of human faces in order to gain access to a computer or computer network: *their site uses Passface because it's less hackable than regular passwords.*

2 (**pass•face**) a digital photograph of a human face that is used for identification in a Passface system: *the company uses cameras and passfaces to make sure only authorized employees get through the door.*

ORIGIN on the pattern of *password.*

pass•word /'pæsˌwərd/ ▸ **n.** a string of characters that allows someone access to a computer system.

paste /pāst/ ▸ **v.** [trans.] insert (a copy of text, graphics, etc.) into a document. See also **CUT AND PASTE.**

ORIGIN late Middle English: from Old French, from late Latin *pasta* 'medicinal preparation in the shape of a small square,' probably from Greek *pastē*, (plural) *pasta* 'barley porridge,' from *pastos* 'sprinkled.'

patch /pæCH/ ▸ **n.** a small piece of code inserted into a program to improve its functioning or to correct a fault: *the patch fixes several graphics bugs.*

▸ **v.** [trans.] correct, enhance, or modify (a routine or program) by inserting a patch.

ORIGIN late Middle English: perhaps from a variant of Old French *pieche*, dialect variant of *piece* 'piece.'

patch•board /ˈpæCH,bôrd/ ▸ **n.** another term for **PATCH PANEL**.

patch pan•el /ˈpæCH ˌpænl/ (also **patch•board**) ▸ **n.** a board in a computer or other device with a number of electric sockets that can be connected in various combinations.

path /pæTH/ ▸ **n.** (pl. **paths** /pæTHz; pæTHs/) a definition of the order in which an operating system or program searches for a file or executable program.

DERIVATIVES **path•less adj.**

ORIGIN Old English *pæth*, of West Germanic origin; related to Dutch *pad*, German *Pfad*, of unknown ultimate origin.

path•name /ˈpæTH,nām/ (also **path name**) ▸ **n.** a description of where a file or other item is to be found in a hierarchy of directories.

PC ▸ **abbr.** personal computer.

PC card /ˈpē ˈsē ˌkärd/ ▸ **n.** a printed circuit board for a personal computer, esp. one built to the PCMCIA standard.

PCI ▸ **n.** a standard for connecting computers and their peripherals.

ORIGIN late 20th cent.: abbreviation of *Peripheral Component Interconnect.*

PCM ▸ **abbr.** pulse code modulation.

PCMCIA ▸ **abbr.** denoting a standard specification for memory cards and interfaces in personal computers: *a PCMCIA card.*

ORIGIN late 20th cent. abbreviation of *Personal Computer Memory Card International Association.*

PC mod /ˈpē ˈsē ˈmäd/ ▸ **n.** another term for **MOD**.

p-code /ˈpē ˌkōd/ ▸ **n.** another term for **PSEUDOCODE.**

PDA ▸ **n.** a handheld computer used to store information such as addresses and telephone numbers, and that can perform many of the same tasks as a personal computer.

ORIGIN late 20th cent.: abbreviation of *personal digital assistant.*

PDF ▸ **n.** a file format that provides an electronic image of text or text and graphics that looks like a printed document and can be viewed, printed, and electronically tramsmitted.

ORIGIN abbreviation of *portable document format.*

peek /pēk/ ▸ **n.** (usu. **PEEK**) a statement or function in BASIC for reading the contents of a specified memory location. Compare with **POKE.**

ORIGIN late Middle English *pike, pyke,* of unknown origin.

peer-to-peer /ˈpi(ə)r tə ˈpi(ə)r/ ▸ **adj.** [attrib.] denoting computer networks in which each computer can act as a server for the others, allowing shared access to files and peripherals without the need for a central server.

pen /pen/ ▸ **n.** an electronic penlike device used in conjunction with a writing surface to enter commands or data into a computer.

ORIGIN Middle English (originally denoting a feather with a sharpened quill): from Old French *penne,* from Latin *penna* 'feather' (in late Latin 'pen').

pe•riph•er•al /pəˈrifərəl/ ▸ **n.** a device that can be attached to and used with a computer.

▸ **adj.** [attrib.] (of a device) able to be attached to and used with a computer, although not an integral part of it.

Perl /pərl/ ▸ **n.** a versatile computer programming language used to process text and add functionality to Web pages.

ORIGIN possibly an acronym of "practical extraction and report language."

per•son•al com•pu•ter /ˈpərsənl kəmˈpyo͞otər/ ▸ **n.** a desktop or notebook computer designed for personal use.

per•son•al dig•i•tal as•sis•tant /ˈpərsənl ˈdijitl əˈsistənt/ ▸ **n.** see **PDA.**

per•son•al in•for•ma•tion man•age•ment /ˈpərsənl ˌinfərˈmāsʜən ˌmænijmənt/ (abbr.: **PIM**) ▸ **n.** the organization of a user's personal information in a program package that includes an address book, organizer, calendar, etc.: *many people buy PDAs for personal information management.*

DERIVATIVES **per•son•al in•for•ma•tion man•ag•er** n.

PIM ▸ abbr. personal information management (or manager).

pipe /pīp/ ▸ n. **1** a command that causes the output from one routine to be the input for another: *you can use a pipe to send the output of the show command to the printer directly and then remove the messages after printing.*

2 a connection to the Internet or to a Web site: *although many businesses have high-powered pipes, the vast majority of home users still have to dial up and wait a seeming eternity for Web pages to pop up.*

ORIGIN short for *pipeline.*

pipe•line /ˈpīpˌlīn/ ▸ n. a linear sequence of specialized modules used for pipelining.

▸ v. [trans.] [often as adjective] (**pipelined**) design or execute (a computer or instruction) using the technique of pipelining.

pipe•lin•ing /ˈpīpˌlīniNG/ ▸ n. a form of computer organization in which successive steps of an instruction sequence are executed in turn by a sequence of modules able to operate concurrently, so that another instruction can be begun before the previous one is finished.

pitch /piCH/ ▸ n. the density of typed or printed characters on a line, typically expressed as numbers of characters per inch.

pix•el /ˈpiksəl/ ▸ n. a minute area of illumination on a display screen, one of many from which an image is composed.

ORIGIN 1960s: abbreviation of *picture element.*

PL ▸ abbr. programming language.

place•hold•er /ˈplāsˌhōldər/ ▸ n. a symbol or piece of text used in an instruction in a computer program to denote a missing quantity or operator.

plat•form /ˈplætˌfôrm/ ▸ n. the particular pairing of a processor and an operating system, which affect how application software must be written in order to run on the computer: *either the Macintosh or Windows platform.*

ORIGIN mid 16th cent.: from French *plateforme* 'ground plan,' literally 'flat shape.'

plat•form game /ˈplætˌfôrm ˌgām/ ▸ **n.** a type of video game featuring two-dimensional graphics in which the player controls a character jumping or climbing between solid platforms at different positions on the screen.

plat•ter /ˈplætər/ ▸ **n.** a rigid rotating disk on which data is stored in a disk drive; a hard disk (considered as a physical object).

ORIGIN Middle English: from Anglo-Norman French *plater*, from *plat* 'large dish'.

Plug and Play /ˈpləg ən(d) ˈplā/ (also **plug and play**) ▸ **n.** a standard for the connection of peripherals to personal computers, whereby a device only needs to be connected to a computer in order to be configured to work properly, without any action by the user: [as adjective] *plug-and-play circuit boards.*

plug-com•pat•i•ble /ˈpləg kəmˌpætəbəl/ ▸ **adj.** relating to or denoting computing equipment that is compatible with devices or systems produced by different manufacturers, to the extent that it can be plugged in and operated successfully.

▸ **n.** a piece of computing equipment designed in this way.

plug-in /ˈpləg ˌin/ ▸ **adj.** (of a module or software) able to be added to a system to give extra features or functions: *a plug-in graphics card.*

▸ **n.** a module or software of this kind: *a plug-in that turns your Web browser into a PDF viewer.*

ply /plī/ ▸ **n.** (pl. **plies**) (in game theory) the number of levels at which branching occurs in a tree of possible outcomes, typically corresponding to the number of moves ahead (in chess strictly half-moves ahead) considered by a computer program.

■ a half-move (i.e., one player's move) in computer chess.

ORIGIN late Middle English: from French *pli* 'fold,' from the verb *plier*, from Latin *plicare* 'to fold.'

PNP ▸ **abbr.** Plug and Play.

point-and-click /ˈpoint ən(d) ˈklik/ ▸ **adj.** (of an interface) giving the user the ability to initiate tasks by using a mouse to move a cursor over an area of the screen and clicking on it.

▸ **v.** [intrans.] use a mouse in such a way.

point•er /ˈpointər/ ▸ **n. 1** another term for **CURSOR**.
2 a variable whose value is the address of another variable; a link.

point•ing de•vice /ˈpointiNG diˌvīs/ ▸ **n.** a generic term for any device (e.g., a graphics tablet, mouse, stylus, or trackball) used to control the movement of a cursor on a computer screen.

poke /pōk/ ▸ **n.** (usu. **POKE**) a statement or function in BASIC for altering the contents of a specified memory location. Compare with **PEEK.**
ORIGIN Middle English: origin uncertain; compare with Middle Dutch and Middle Low German *poken*, of unknown ultimate origin.

Po•lish no•ta•tion /ˈpōlisH nōˈtāsHən/ ▸ **n.** Logic a system of formula notation without brackets or special punctuation, frequently used to represent the order in which arithmetical operations are performed in many computers and calculators. In the usual form (**reverse Polish notation**), operators follow rather than precede their operands.

poll /pōl/ ▸ **v.** [trans.] check the status of (a measuring device, part of a computer, or a node in a network), esp. as part of a repeated cycle.

pol•y•mor•phism /ˌpäliˈmôrˌfizəm/ ▸ **n.** a feature of a programming language that allows routines to use variables of different types at different times.
DERIVATIVES **pol•y•mor•phic** /-ˈmôrfik/ **adj.**; **pol•y•mor•phous** /-ˈmôrfəs/ **adj.**

pol•y•no•mi•al time /ˌpäliˈnōmēəl ˌtīm/ ▸ **n.** the time required for a computer to solve a problem, where this time is a simple polynomial function of the size of the input.

POP /päp/ (also **PoP**) ▸ **abbr.** point of presence, an Internet access point, such as that provided by an ISP, with a unique IP address.

POP3 /ˈpäp ˈTHrē/ ▸ **n.** a protocol for receiving e-mail by downloading it to your computer from a mailbox on the server of an Internet service provider.
ORIGIN abbreviation of *Post Office Protocol 3*

pop-un•der /ˈpäp ˌəndər/ ▸ **adj.** of, relating to, or denoting an additional window, usually an advertisement, that is under a Web browser's main or current window and appears when a user tries to exit: *a plague of flashing pop-under ads.*

pop-up /ˈpäp ˌəp/ ▸ **adj.** **1** denoting Internet advertisements in their own windows which appear adventitiously when visiting Web sites or following links: *a U.S. district court judge has ruled that online advertising companies can launch pop-up ads on other parties' Web sites, the Wall Street Journal reported.*

2 denoting a running program that is activated by one or more keystrokes: *users can control the buffer switch via pop-up memory-resident software and can thus direct their output to one of several printers.*

▸**n.** a pop-up ad, window, or program.

port[1] /pôrt/ ▸ **n.** a socket in a computer into which a device can be plugged: *a USB port.*

ORIGIN Old English: from Latin *porta* 'gate'; reinforced in Middle English by Old French *porte*. The later sense 'opening in the side of a ship' led to the general sense 'aperture.'

port[2] /pôrt/ ▸ **v.** [trans.] transfer (software) from one system or machine to another: *they ported their IBM supercomputing software to Linux-based software servers.*

▸ **n.** a transfer of software from one system or machine to another.

ORIGIN Middle English: from Old French *port* 'bearing, gait,' from the verb *porter*, from Latin *portare* 'carry.' The verb (from French *porter*) dates from the mid 16th cent.

port•a•ble /ˈpôrṯəbəl/ ▸ **n.** a small computer that can be easily carried.

▸ **adj.** (of software) able to be transferred from one machine or system to another.

DERIVATIVES **port•a•bil•i•ty n.**; **port•a•bly adv.**

ORIGIN late Middle English: from Old French *portable*, from late Latin *portabilis*, from Latin *portare* 'carry.'

por•tal /ˈpôrtl/ ▸ **n.** an Internet site providing access or links other sites: *there is a name for this sort of Web service: a portal. The term signifies an Internet gateway, the shoreline from which Web surfers paddle out to sea.*

Posix /ˈpäsiks/ (also **POSIX**) ▸ **n.** a set of formal descriptions that provide a standard for the design of operating systems, esp. ones that are compatible with Unix.

ORIGIN 1980s: from the initial letters of *portable operating system* + *-ix* suggested by **UNIX**.

post /pōst/ ▸ **v.** [trans.] (often **be posted**) make (information) available on the Internet.

ORIGIN Old English, from Latin *postis* 'doorpost,' later 'rod, beam,' probably reinforced in Middle English by Old French *post* 'pillar, beam' and Middle Dutch, Middle Low German *post* 'doorpost.'

post•er /ˈpōstər/ ▸ **n.** someone who sends a message to a newsgroup.

post•er•ize /ˈpōstəˌrīz/ ▸ **v.** [trans.] print or display (a photograph or other image) using only a small number of different tones.

DERIVATIVES **post•er•i•za•tion n.**

post•ing /ˌpōstiNG/ ▸ **n.** a message sent to an Internet bulletin board or newsgroup: *tap into the postings in the Internet and you're swamped by a maelstrom of communicative debris.*

■ the action of sending a message to an Internet bulletin board or newsgroup.

Post•Script /ˈpōstˌskript/ ▸ **n.** trademark a language used as a standard for formatting pages of text: [mainly as modifier] *Create your document and print to PostScript file. Then Distill your PostScript file to create your PDF document to be printed for public distribution.*

pow•er rat•ing /ˈpow-ər ˌrātiNG/ ▸ **n.** the amount of electrical power required for a particular device: *a continuous power rating of 150 watts.*

pow•er-up /ˈpow-ər ˌəp/ ▸ **n.** the action of switching on a computer.

■ (in a computer game) a bonus that a player can collect and that gives their character an advantage, such as more strength or firepower.

pow•er us•er /ˈpow-ər ˌyo͞ozər/ ▸ **n.** an accomplished computer user who requires products having the most features and the fastest performance: *the Gateway 700XL's record-breaking performance packed into a stylish, easy-to-upgrade case is sure to please power users.*

ppi ▸ **abbr.** pixels per inch, a measure of the resolution of display screens, scanners, and printers.

ppm ▸ **abbr.** page(s) per minute, a measure of the speed of printers.

PPP ▸ abbr. point to point protocol, a method by which a computer communicates over a telephone line or fiber optic cable with a server providing a connection to the Internet.

pre•fetch ▸ v. /prē'fecH/ [trans.] transfer (data) from main memory to temporary storage in readiness for later use.
▸ n. /'prē͵fecH/ a process involving such a transfer.

pre•in•stall /͵prēin'stôl/ ▸ v. another term for PRELOAD.

pre•load /prē'lōd/ ▸ v. [trans.] load beforehand: *the software comes preloaded on the PC.*
▸ n. something loaded or applied as a load beforehand: *prices include DOS and Windows preload.*

pre•mas•ter /prē'mæstər/ ▸ v. [trans.] make a master copy of (data) on a hard disk before writing it to a CD or DVD.

pre•proc•es•sor /prē'präsesər/ ▸ n. a computer program that modifies data to conform with the input requirements of another program.

pre•pro•gram /prē'prō͵græm/ ▸ v. [trans.] [usu. as adjective] (**preprogrammed**) program (a computer or other electronic device) in advance for ease of use: *a preprogrammed function key.*
■ program (something) into a computer or other electronic device before use: *preprogrammed messages.*

pres•en•ta•tion graph•ics /͵prezən'tāsHən ͵græfiks; ͵prēzen-/ ▸ n. another term for PRESENTATION SOFTWARE.

pres•en•ta•tion soft•ware /͵prezən'tāsHən ͵sôftwe(ə)r; ͵prēzen-/ (also **pres•en•ta•tion graph•ics** /͵prezən'tāsHən ͵grafiks; ͵prēzen-/) ▸ n. software used to create a sequence of text and graphics, and often audio and video, to accompany a speech or public presentation: *the suite comes complete with a word-processor, spreadsheet, presentation software and various other components.*

pre•view /'prē͵vyo͞o/ ▸ n. a facility in a software program for inspecting the appearance of a document before it is printed.
▸ v. [trans.] see or inspect (a document) before it is printed.

prim•i•tive /'primitiv/ ▸ n. a simple operation or procedure of a limited set from which complex operations or procedures may be constructed, esp. a simple geometric shape that may be generated in computer graphics by such an operation or procedure.

ORIGIN late Middle English: from Old French *primitif, -ive*, from Latin *primitivus* 'first of its kind,' from *primus* 'first.'

print /print/ ▸ **v.** [trans.] (often **be printed**) produce a paper copy of (information stored on a computer): *the results of a search can be printed out.*

ORIGIN Middle English (denoting the impression made by a stamp or seal): from Old French *preinte* 'pressed,' feminine past participle of *preindre*, from Latin *premere* 'to press.'

print•a•ble /ˈprintəbəl/ ▸ **adj.** able to be printed: *software with downloadable songs and printable sheet music.*

print•er /ˈprintər/ ▸ **n.** a device linked to a computer for printing text or pictures onto paper.

print•er-friend•ly /ˈprintər ˌfrendlē/ ▸ **adj.** formatted for output on a printer, with extraneous material deleted or suppressed: *users can also print printer-friendly sample ballots to take with them to their polling place.*

print•head /ˈprintˌhed/ (also **print head**) ▸ **n.** a component in a printer that assembles and holds the characters and from which the images of the characters are transferred to the printing medium.

print•out /ˈprintˌowt/ ▸ **n.** a page or set of pages of printed material produced by a computer's printer.

print queue /ˈprint ˌkyo͞o/ ▸ **n.** a series of print jobs waiting for a printer to be ready to print them.

pri•vate key /ˈprīvit ˈkē/ ▸ **n.** a cryptographic key known only to the sender who encrypted a message and to the recipient who will decrypt it, used as the only key or in addition to a **public key**. See also **ONE-TIME PAD**.

pro•ce•dure /prəˈsējər/ ▸ **n.** another term for **SUBROUTINE**.

ORIGIN late 16th cent.: from French *procédure*, from *procéder*.

proc•ess /ˈpräses/ ▸ **v.** operate on (data) by means of a program.

ORIGIN Middle English: from Old French *proces*, from Latin *processus* 'progression, course,' from the verb *procedere*.

proc•es•sor /ˈpräsesər/ ▸ **n.** an integrated circuit that contains the computer's CPU.

pro•gram /ˈprōˌgræm/ ▸ **n.** a series of coded software instructions to control the operation of a computer or other machine.

▸ v. (**pro•grammed, pro•gram•ming**; or **pro•gramed, pro•gram•ing**) [trans.] provide (a computer or other machine) with coded instructions for the automatic performance of a particular task: *it is a simple matter to program the computer to recognize such symbols.*
■ input (instructions for the automatic performance of a task) into a computer or other machine: *simply* ***program in*** *your desired volume level.*
DERIVATIVES **pro•gram•ma•ble** /ˈprō,græməbəl/ **adj.**

pro•gram•mer /ˈprō,græmər/ ▸ **n.** a person who writes computer programs.

pro•gram•ming /ˈprō,græmiNG/ ▸ **n.** the action or process of writing computer programs.

Pro•log /ˈprō,lôg/ ▸ **n.** a high-level computer programming language first devised for artificial intelligence applications.
ORIGIN 1970s: from the first elements of **PROGRAMMING** and **LOGIC**.

PROM /präm/ ▸ **n.** a memory chip that can be programmed only once by the manufacturer or user.
ORIGIN from *p(rogrammable) r(ead-)o(nly) m(emory).*

prompt /prämpt/ ▸ **v.** [trans.] (of a computer) request input from (a user).
▸ **n.** a word or symbol on a monitor to show that the system is waiting for input.
ORIGIN Middle English (as a verb): based on Old French *prompt* or Latin *promptus* 'brought to light,' also 'prepared, ready,' past participle of *promere* 'to produce,' from *pro-* 'out, forth' + *emere* 'take.'

pro•pel•ler-head /prəˈpelər ,hed/ ▸ **n.** informal a person who has an obsessive interest in computers or technology.

pro•tect /prəˈtekt/ ▸ **v.** [trans.] restrict access to or use of (data or a memory location): *security products are designed to protect information from unauthorized access.*
DERIVATIVES **pro•tect•a•ble adj.**
ORIGIN late Middle English: from Latin *protect-* 'covered in front,' from the verb *protegere*, from *pro-* 'in front' + *tegere* 'to cover.'

pro•to•col /ˈprōt̲ə,kôl/ ▸ **n.** a set of rules governing the exchange or transmission of data electronically between devices.

ORIGIN late Middle English (denoting the original record of an agreement, forming the legal authority for future dealings relating to it): from Old French *prothocole*, via medieval Latin from Greek *prōtokollon* 'first page, flyleaf,' from *prōtos* 'first' + *kolla* 'glue.'.

pseu•do•code /'so͞odōˌkōd/ ▸ **n.** a notation resembling a simplified programming language, used in program design.

pseu•do•ran•dom /ˌso͞odō'rændəm/ ▸ **adj.** (of a number, a sequence of numbers, or any digital data) satisfying one or more statistical tests for randomness but produced by a definite mathematical procedure.

DERIVATIVES **pseu•do•ran•dom•ly adv.**

pub•lic key /'pəblik 'kē/ ▸ **n.** a cryptographic key that can be obtained and used by anyone to encrypt messages intended for a particular recipient, such that the encrypted messages can be deciphered only by using a second key that is known only to the recipient (the **private key**).

puck /pək/ ▸ **n.** an input device similar to a mouse that is dragged across a sensitive surface, which notes the puck's position to move the cursor on the screen.

pull-down /'po͝ol ˌdown/ ▸ **adj.** [attrib.] (of a menu) appearing below a menu title only while selected. Compare with **DROP-DOWN**.

▸ **n.** a pull-down menu.

punched card /'pənCHt 'kärd/ (also **punch card** /'pənCH ˌkärd/) ▸ **n.** a card perforated according to a code, for controlling the operation of a machine, used in voting machines and formerly in programming computers.

punched tape /'pənCHt 'tāp/ ▸ **n.** a paper tape perforated according to a code, formerly used for conveying instructions or data to a computer.

push /po͝oSH/ ▸ **n.** automatic transmission of information over the Internet from a server to a client program on a computer, without a user request but based on the user's previously specified preferences.

▸ **v.** [trans.] prepare (a stack) to receive a piece of data on the top.
■ transfer (data) to the top of a stack.

puz•zler /ˈpəz(ə)lər/ ▸ **n.** informal a computer game in which the player must solve puzzles.

Python /ˈpīTHän/ ▸ **n.** a computer programming language used to build Web sites.

Q

quan•tum com•put•er /ˈkwäntəm kəmˌpyo͞otər/ ▸**n.** a computer that makes use of the quantum states of subatomic particles to store information: *in 1992 Deutsch and Richard Jozsa formulated a few problems that could be solved faster with a quantum computer than with a conventional Turing machine.*

quan•tum com•put•ing /ˈkwäntəm kəmˌpyo͞ot̲iNG/ ▸**n.** a theoretical computing model based on quantum theory and the behavior of subatomic particles, potentially providing computing power far in excess of the largest supercomputers.

que•ry lan•guage /ˈkwi(ə)rē ˌlæNGgwij/ ▸**n.** a language for the specification of procedures for the retrieval (and sometimes also modification) of information from a database.

queue /kyo͞o/ ▸**n.** a list of data items, commands, system requests, etc., stored so as to be retrievable in a definite order, usually the order of insertion: *when you click Send, the message goes into a queue.*
▸**v.** (**queues**, **queued**, **queu•ing** or **queue•ing**) [trans.] arrange in a queue.
ORIGIN late 16th cent. (as a heraldic term denoting the tail of an animal): from French, based on Latin *cauda* 'tail.'

qwerty /ˈkwərt̲ē/ ▸**adj.** denoting the standard layout on English-language keyboards, having *q*, *w*, *e*, *r*, *t*, and *y* as the first keys from the left on the top row of letters.

ra•di•o but•ton /ˈrādēō ˌbətn/ ▸ **n.** a small circle beside each item in a list, only one of which may be chosen, that is filled with a black dot when selected by the user.

RAID /rād/ ▸ **abbr.** redundant array of independent (or inexpensive) disks, a system for providing greater capacity, faster access, and security against data corruption by spreading the data across several disk drives.

RAM /ræm/ ▸ **abbr.** random-access memory.

ran•dom ac•cess /ˈrændəm ˈækˌses/ ▸ **n.** the process of transferring information to or from memory in which every memory location can be accessed directly rather than being accessed in a fixed sequence: [as adjective] *random-access programming.*

ran•dom•ize /ˈrændəˌmīz/ ▸ **v.** [trans.] [usu. as adjective] (**randomized**) make unpredictable, unsystematic, or random in order or arrangement; employ random selection or sampling in (a program or procedure).

DERIVATIVES **ran•dom•i•za•tion n.**

ras•ter /ˈræstər/ ▸ **n.** the series of vertical lines of pixels that form the image on a CRT monitor.

■ the area the image fills on the screen of a CRT monitor, which is a slightly smaller area then the screen itself.

ras•ter im•age proc•es•sor /ˈræstər ˈimij ˌpräsesər/ (abbr.: **RIP**) ▸ **n.** a device that rasterizes an image.

ras•ter•ize /ˈræstəˌrīz/ ▸ **v.** [trans.] convert (an image stored as an outline) into pixels that can be displayed on a screen or printed.

DERIVATIVES **ras•ter•i•za•tion** /ˌræstərəˈzāsʜən/ **n.**; **ras•ter•iz•er n.**

RDBMS ▸ **abbr.** relational database management system.

read /rēd/ ▸ **v.** (past and past part. **read** /red/) [trans.] (of a computer) copy or transfer (data).

■ enter or extract (data) in an electronic storage device: *the commonest way of reading a file into the system.*

ORIGIN Old English *rǣdan*, of Germanic origin; related to Dutch *raden* and German *raten* 'advise, guess.' Early senses included 'advise' and 'interpret (a riddle or dream)'.

read•er /ˈrēdər/ ▸ **n.** a device or piece of software used for reading or obtaining data from a storage medium.

ORIGIN Old English *rǣdere* 'interpreter of dreams, reader.'

read-in /ˈrēd ˌin/ ▸ **n.** the input or entry of data to a computer or storage device.

read-on•ly mem•o•ry /ˈrēd ˌōnlē ˈmem(ə)rē/ (abbr.: **ROM**) ▸ **n.** memory read at high speed but not able to be changed or deleted by program instructions or user input.

read•out /ˈrēdˌowt/ (also **read-out**) ▸ **n.** a visual record or display of the output from a computer.

■ the process of transferring or displaying such data.

read/write /ˈrēd ˈrīt/ (also **read-write**) ▸ **adj.** denoting functions related to both reading and writing data to computer memory or storage: *read/write operations.*

■ able to both read data from and write data to a storage medium such as a disk: *I am having a problem using my read-write CD drive.*

real time /ˈrē(ə)l ˌtīm/ ▸ **n.** of or relating to a system in which input data is processed within milliseconds so that it is available virtually immediately as feedback: *real-time communications, such as instant messaging.*

re•boot ▸ **v.** /rēˈbo͞ot/ [trans.] boot (a computer system) again.

■ [intrans.] (of a computer system) be booted again.

▸ **n.** /ˈrē ˌbo͞ot; rēˈbo͞ot/ an act or instance of booting a computer system again.

re•call ▸ **v.** /riˈkôl/ [trans.] call up (stored computer data) for processing or display.

▸ **n.** /ˈrēˌkôl/ the proportion of the number of relevant documents retrieved from a database in response to an inquiry.

ORIGIN late 16th cent. (as a verb): from *re-* 'again' + *call*, suggested by Latin *revocare* or French *rappeler* 'call back.'

re•code /rē'kōd/ ▸ **v.** [trans.] put (something, esp. a computer program) into a different code.

■ assign a different code to (data).

rec•og•ni•tion /ˌrekəg'nisHən/ ▸ **n. 1** the process of recognizing data input to the computer in a particular form: *there's some fancy software for handwriting recognition* **2** speech recognition technology.

ORIGIN late 15th cent. (denoting the acknowledgment of a service): from Latin *recognitio(n-)*, from the verb *recognoscere* 'know again, recall to mind' (see **RECOGNIZE**).

rec•og•nize /'rekəgˌnīz/ ▸ **v.** [trans.] (of a computer) automatically identify and respond correctly to (a sound, picture, printed character, etc.).

■ (of a computer) identify and load a driver to operate a new device attached to the computer: *the system failed to recognize the LCD monitor.*

ORIGIN late Middle English (earliest attested as a term in Scots law): from Old French *reconniss-*, stem of *reconnaistre*, from Latin *recognoscere* 'know again, recall to mind,' from *re-* 'again' + *cognoscere* 'learn.'

re•com•pile /ˌrēkəm'pīl/ ▸ **v.** [trans.] compile (a program) again or differently.

▸ **n.** a recompilation of a computer program: *users should upgrade to SendMail version 8.12.10 or apply the provided patch. (This requires a recompile.)*

DERIVATIVES **re•com•pi•la•tion** /ˌrēkämpə'lāsHən/ **n.**

rec•ord /'rekərd/ ▸ **n.** a number of related items of information that are handled as a unit.

ORIGIN Middle English: from Old French *record* 'remembrance,' from *recorder* 'bring to remembrance,' from Latin *recordari* 'remember,' based on *cor, cord-* 'heart.'

rec•ord•set /'rekərdˌset/ ▸ **n.** a set of records in a database that share an identifiable or isolatable characteristic.

re•cur•sive /ri'kərsiv/ ▸ **adj.** relating to or involving a program or routine of which a part requires the application of the whole, so that its explicit interpretation requires in general many successive executions.

DERIVATIVES **re•cur•sive•ly** adv.

re•for•mat /rē'fôr,mæt/ ▸ v. (**re•for•mat•ted**, **re•for•mat•ting**) [trans.] give a new format to (a data storage medium): *reformat the hard disk.*

■ revise or represent (data) in another format.

re•fresh /ri'fresH/ ▸ v. [trans.] update the display on (a monitor screen).

■ (also **re•load**) update the display of a Web page on a monitor screen. ■ update dynamic random-access memory.

▸ n. an act or function of updating the display on a monitor screen.

ORIGIN late Middle English: from Old French *refreschier*, from *re-* 'back' + *fres(che)* 'fresh.'

re•fresh rate /ri'fresH ,rāt/ ▸ n. the number of times per second that a monitor's display is updated, measured in hertz.

reg•is•ter /'rejistər/ ▸ n. (in electronic devices) a location in a store of data, used for a specific purpose and with quick access time. ▸ v. [trans.] to record one's name and serial number of a software product with the product's manufacturer in order to be able to receive technical support or receive updates.

ORIGIN late Middle English: from Old French *regestre* or medieval *regestrum*, *registrum*, alteration of *regestum*, singular of late Latin *regesta* 'things recorded,' from *regerere* 'enter, record.'

re•in•stall /,rē-in'stôl/ ▸ v. [trans.] install again (used especially of software).

▸ n. a reinstallation of software: *if the preceding two tips don't help, try performing a "clean" reinstall of your system software.*

DERIVATIVES **re•in•stal•la•tion** /,rē-instə'lāsHən/ n.

re•key /rē'kē/ ▸ v. [trans.] enter (text or other data) again using a keyboard.

rela•tion•al da•ta•base /ri'lāsHənl 'dætə,bās; 'dātə,bās/ ▸ n. a database structured to recognize relations between stored items of information.

re•mail•er /'rē,mālər/ ▸ n. a service that anonymously forwards e-mail so as to disguise the original sender: *the need for anony-*

mous remailers stems from the design of the Internet, which tags every packet of data with an electronic address so it can be returned or re-sent if something goes wrong in transit.

re•map /ˌrē'mæp/ ▸ v. (**re•mapped**, **re•map•ping**) [trans.] assign (a function) to a different key.

re•mote /ri'mōt/ ▸ adj. (**re•mot•er**, **re•mot•est**) denoting a device that can only be accessed by means of a network. Compare with **LOCAL**.
DERIVATIVES **re•mote•ly** [adv.]
ORIGIN late Middle English: from Latin *remotus* 'removed,' past participle of *removere*.

rend•er /'rendər/ ▸ v. [trans.] process (an outline image) using color and shading in order to make it appear solid and three-dimensional.

rend•er•ing /'rendəriNG/ ▸ n. the processing of an outline image using color and shading to make it appear solid and three-dimensional: *3D graphics rendering technology.*

re•par•ti•tion /ˌrēpär'tisHən/ ▸ v. [trans.] divide (a hard disk) into partitions again: *the restore disk has everything you need to repartition the drive, format, and reload software to its original condition.*

re•pur•pose /rē'pərpəs/ ▸ v. [trans.] adapt for use in a different purpose: *they've taken a product that was originally designed for a CD-ROM and repurposed it for the Microsoft Network.*

re•quest /ri'kwest/ ▸ n. an instruction to a computer to provide information or perform another function.
ORIGIN Middle English: from Old French *requeste* (noun), based on Latin *requirere*.

re•served word /ri'zərvd 'wərd/ ▸ n. a word in a programming language that has a fixed meaning and cannot be redefined by the programmer.

res•i•dent /'rezidənt/ ▸ adj. (of a computer program, file, etc.) immediately available in computer memory, rather than having to be loaded from elsewhere.
ORIGIN Middle English: from Latin *resident-* 'remaining,' from the verb *residere*.

re•size /rē'sīz/ ▸ v. [trans.] alter the size of (a computer window, image, etc.): *the editing program lets you rotate, crop, and resize images.*

res•o•lu•tion /ˌrezə'lo͞osHən/ ▸ n. the degree of detail visible on computer monitor, measured as the number of pixels horizontally and vertically.

re•sponse time /ri'späns ˌtīm/ ▸ **n.** the length of time taken for a computer system to respond to a user request or inquiry or command.

ret•i•nal scan•ner /'retn-əl 'skænər/ ▸ **n.** a biometric device that scans a person's or animal's retina in infrared for identification purposes: *when I worked at Winstar, someone from another department set up a mantrap with a retinal scanner to protect the tech center.*

re•triev•al /ri'trēvəl/ ▸ **n.** the obtaining or consulting of material stored in a computer system.

re•trieve /ri'trēv/ ▸ **v.** [trans.] find or extract (information stored in a computer).

DERIVATIVES **re•triev•a•bil•i•ty n.**; **re•triev•a•ble adj.**

re•try ▸ **v.** /rē'trī/ (**re•tries**, **re•tried**) [intrans.] reenter a command, esp. differently because one has made an error the first time.

■ transmit data again because the first attempt was unsuccessful.

▸ **n.** /'rēˌtrī/ an instance of reentering a command or retransmitting data.

re•turn /ri'tərn/ ▸ **n.** (also **re•turn key**) on some computer keyboards, a key that is used to perform various functions, such as executing a command or selecting options on a menu.

ORIGIN Middle English: the verb from Old French *returner*, from Latin *re-* 'back' + *tornare* 'to turn'; the noun via Anglo-Norman French.

re•verse Po•lish no•ta•tion /ri'vərs 'pōlisн nō'tāsнən/ ▸ **n.** see **POLISH NOTATION.**

re•writ•able /rē'rītəbəl/ ▸ **adj.** (of digital storage media) capable of being written to and erased multiple times: *the subsidiary's first products are a 650-megabyte rewritable, removable optical drive and a 4mm DAT tape drive.*

RFC ▸ **abbr.** request for comment, a document from the Internet Engineering Task Force, which is responsible for defining Internet protocols, that forms the basis of a technical standard.

right-click /'rīt 'klik/ ▸ **v.** [intrans.] depress the right-hand button on a mouse.

■ [trans.] click on a link or other screen object in this way: *right-click a graphic and choose Resize.*

▸**n.** the action of right-clicking: [as modifier] *right-click features.*

rip /rip/ ▸**v.** (**ripped, rip•ping**) [trans.] use a program to copy (a sound sequence on a compact disc) on to a computer's hard drive. *I have every Beatles song ever made (ripped from my boxed set of CDs at a bit-rate of 192).*

RIP /rip/ ▸**abbr.** raster image processor.

▸**v.** (usu. **rip**) (**ripped, rip•ping**) [trans.] rasterize (an image): *once you are happy with the image, you can rip it out.*

RISC /risk/ ▸**n.** [usu. as adjective] a computer based on a processor or processors designed to perform a limited set of operations extremely quickly: *RISC-based systems.*

■ computing using this kind of computer.

ORIGIN 1980s: acronym from *reduced instruction set computer* (or *computing*).

roam /rōm/ ▸**v.** [trans.] move from site to site on (the Internet): *roam the Internet.*

■ use wireless network access in locations operated by different service providers: *wireless roaming* | *roaming across public wireless hotspots.*

ORIGIN Middle English: of unknown origin.

role-play•ing game /ˈrōl ˌplāiNG ˌgām/ ▸**n.** a game in which players take on the roles of imaginary characters who engage in adventures, typically in a particular computerized fantasy setting overseen by a referee.

roll•back /ˈrōlˌbæk/ ▸**n.** the process of restoring a database or program to a previously defined state, typically to recover from an error.

▸**v.** [trans.] restore (a database) to a previously defined state.

roll•er•ball /ˈrōlərˌbôl/ ▸**n.** an input device containing a ball that is moved with the fingers to control the cursor.

roll-up /ˈrōl ˌəp/ ▸**adj.** [attrib.] denoting a menu that will display only its title to save screen space.

ROM /räm/ ▸**abbr.** read-only memory.

ROM•ve•lope /ˈrämvəˌlōp/ (also **rom•ve•lope**) ▸**n.** a protective envelope or sleeve, usually made of cardboard, used to package or mail a compact disc: *you can have one of these ROMvelopes to protect your CD.*

ORIGIN blend of *(CD-)ROM* and *(en)velope.*

root di•rec•to•ry /ˈro͞ot dəˌrektərē/ ▸ **n.** the directory at the highest level of a hierarchy.

rout•er /ˈrowt̲ər; ˈro͞ot̲ər/ ▸ **n.** a device that forwards data packets to the appropriate parts of a computer network: *a router to network together our home computers.*

rou•tine /ro͞oˈtēn/ ▸ **n.** a sequence of instructions for performing a task that forms a program or a distinct part of one.

ORIGIN late 17th cent. (denoting a regular course or procedure): from French, from *route* 'road'.

rout•ing code /ˈro͞otiNG ˌkōd; ˈrowtiNG/ ▸ **n.** any of various codes used to direct data, documents, or merchandise, including:

■ a numeric code that directs telephone calls or Internet traffic.

RPG ▸ **abbr.** ■ role-playing game. ■ report program generator, a high-level commercial computer programming language.

RTF ▸ **abbr.** rich text format, developed to allow the transfer of graphics and formatted text between different applications and operating systems.

RTFM ▸ **abbreviation** vulgar slang read the fucking manual (used especially in e-mail in reply to a question whose answer is patently obvious).

run-time /ˈrən ˌtīm/ ▸ **n.** **1** the length of time a program takes to run.

■ the time at which the program is run: *web services provide referenceable, configurable programming code that can be integrated at run-time to seamlessly produce flexible, dynamic applications.*

2 a cut-down version of a program that can be run but not changed: *you can distribute the run-time to your colleagues.*

▸ **adj.** (of software) in a reduced version that can be run but not changed.

run-time li•cense /ˈrən ˌtīm ˌlīsəns/ ▸ **n.** a relatively broad software license enabling the holder to operate software on a network and in some cases to distribute it with other products.

EIGHT SIMPLE RULES FOR KID-FRIENDLY COMPUTING

There's no shortage of fear, uncertainty, and doubt when it comes to kids and computers. If your children don't have access to a computer, the conventional wisdom goes, they won't learn the skills they need to do well in college and to get a good job. If they do have access to a computer, they are at risk of being exposed to everything from predatory Internet creeps to spam, viruses, and lawsuits from the RIAA.

There are, however, a few common-sense guidelines that you can follow to help make sure your children's computer use is as safe and as rewarding as possible.

1. *Make sure the computer area is set up correctly for little bodies.*
If you share a computer and desk with your children, make sure that they can change the height of the chair or keyboard tray, or even swap out your full-size keyboard and mouse for one made for smaller hands. Make sure your children learn good ergonomic habits early—they should take regular breaks from the computer. (There are quite a few programs you can download that give gentle and regular reminders to take a break.)

2. *Keep the computer where you can keep an eye on both it and your kids.*
Many experts suggest that family computers be kept in open or common areas of the house, not in a child's bedroom. This lets

you be aware of what's on the screen—an IM session? A game? A report for school? Or a questionable website? (FYI: if you see the acronym "POS" on the screen in an IM window as you walk by, it means "Parent Over Shoulder.")

3. *Spend some time using the computer with your children.*
Even if it's only losing to them in their favorite game, spending time together lets you model good computer use. Show them your favorite Web site, or ask them to help you shop online for a present. Check the weather where you are, or where their grandparents are. Look up the lyrics to a song they like. Ask them what their favorite sites are, and visit them together. If they are at home while you are at work, IM them or send them e-mail.

4. *Set computer limits.*
Limit the amount of time your children spend online and on the computer. There are software and hardware components you can purchase that can enforce your limits, if necessary—even ones that will keep the computer from being turned on at all.

5. *Set computer tasks and goals.*
All computer use is not equal. Two hours of chat or IM is not the same as two hours of playing a game, which is not the same as two hours of web searching or two hours of making a multimedia presentation. Make sure that your children are developing well-rounded computer skills by setting goals for them to meet. Set up a web scavenger hunt for them—can they find a Web site devoted to ancient musical instruments? Women's baseball teams? Making kilts? Ask them to format a document five different (readable) ways, or to show the same document as paragraphs, an outline, and a table. Make extra computer privileges dependent on completion of your tasks.

6. *Keep up your own computer skills.*
If you have to ask your ten-year-old for help every time you want to copy a file or make a backup, you won't have very

much authority in guiding her computer use. Make sure you know the basics and keep up with new programs. Know what's on your computer and what it's there for, so you will notice if any new, possibly unsavory, software appears.

7. *Have explicit computer rules.*
Tell your children how you expect them to use the computer. Write it down, and if necessary, ask them to sign a copy. These rules should include any rule you want them to follow from "no drinks by the computer!" to "no downloading music or software without permission." Post a copy by the computer, and emphasize that any friends using your family's computer must also obey the same rules.

8. *Know your child's online friends.*
Know who is sending your child e-mail, and keep an eye on their IM correspondents. Elementary-school children should be receiving e-mail from family and school friends only. Older kids will probably have e-mail from teammates and possibly even kids from around the world who share their interest in a particular band or TV show. Don't just jump in and read their e-mail—ask the same questions you would about school friends or teammates. "What's Amycat1989 like? Where does she live? Does she go to your school?" Be alert for evasive or unlikely answers. Ask to phone their parents, or e-mail their correspondents. Reiterate that your children should never give out their address, phone number, or the name of their school online, and that they should especially never arrange to meet in person anyone they have only known online, without your permission. There are quite a few scary stories out there about Internet predators—every parent has heard them. However, most children can spot a phony, even online, and keeping a watchful eye on your child's activities, online and off, is the best defense.

By guiding and overseeing your children's use of computers, you can keep them safe so that they can learn and enjoy themselves appropriately—and reduce your own stress and worry.

S

save /sāv/ ▸ **n.** an act of copying or moving data to a storage location, usually the hard drive: *creating a new editing context after each save helped; now it goes up to about 13,000 records before slowing down.*

scal•a•ble /ˈskāləbəl/ ▸ **adj.** (of a computing process) able to be used or produced in a range of capabilities: *it is scalable across a range of systems.*

DERIVATIVES **scal•a•bil•i•ty n.**

scale /skāl/ ▸ **v.** [intrans.] change the size of (a graphic).

ORIGIN late Middle English: from Latin *scala* 'ladder' (the verb via Old French *escaler* or medieval Latin *scalare* 'climb'), from the base of Latin *scandere* 'to climb.'

scan /skæn/ ▸ **v.** (**scanned**, **scan•ning**) [trans.] convert (a document or picture) into digital form for storage or processing on a computer: *text and pictures can be scanned into the computer.*

DERIVATIVES **scan•na•ble adj.**

scan•ner /ˈskænər/ ▸ **n.** a device that scans documents and converts them into digital data.

sched•ul•er /ˈskejo͞olər/ ▸ **n.** a program that arranges jobs or a computer's operations into an appropriate sequence.

Scheme /skēm/ ▸ **n.** a computer programming language, an offshoot of Lisp, often used to teach introductory programming.

ORIGIN 1970s: originally Schemer, by analogy with *Planner* and *Conniver*, two other programming languages.

scratch•pad /ˈskræCH,pæd/ (also **scratch pad**) ▸ **n.** a small, fast memory for the temporary storage of data.

screen /skrēn/ ▸ **n.** the surface of a monitor or other electronic device on which images and data are displayed.

■ the data or images displayed on a computer screen: *pressing the F1 key at any time will display a help screen.*

ORIGIN Middle English: shortening of Old Northern French *escren*, of Germanic origin.

screen•ag•er /ˈskrēnˌājər/ ▸ **n.** informal a person in their teens or twenties who has an aptitude for computers and the Internet: *today's "screenager"—the child born into a culture mediated by the television and computer—is interacting with his world in at least as dramatically altered a fashion from his grandfather as the first sighted creature did from his blind ancestors.*

ORIGIN 1990s: blend of *screen* and *teenager*.

screen dump /ˈskrēn ˌdəmp/ ▸ **n.** the process or an instance of causing what is displayed on a monitor screen to be printed out.

■ a resulting printout.

screen sav•er /ˈskrēn ˌsāvər/ (also **screen•sav•er**) ▸ **n.** a program that, after a set time, replaces an unchanging screen display with a moving image to prevent damage to the phosphor of a CRT.

screen shot /ˈskrēn ˌSHät/ (also **screen•shot**) ▸ **n.** a photograph of the display on a computer screen to demonstrate the operation of a program.

script /skript/ ▸ **n.** an automated series of instructions carried out in a specific order: *an e-mail message containing an embedded malicious script.*

ORIGIN late Middle English: shortening of Old French *escript*, from Latin *scriptum*, neuter past participle (used as a noun) of *scribere* 'write.'

script kid•die /ˈskript ˌkidē/ ▸ **n.** informal, derogatory a person who uses existing computer scripts or codes to hack into computers, lacking the expertise to write their own: *he threatened to strike others when they called him a "script kiddie," a know-nothing hacker.*

scroll /skrōl/ ▸ **n.** [usu. as adjective] the facility that moves a display on a computer screen in order to view new material.

▸ **v.** [intrans.] move displayed text or graphics in a particular direction on a computer screen in order to view different parts of them: *she scrolled through her file.*

■ (of displayed text or graphics) move up, down, or across a computer screen.

DERIVATIVES **scroll•a•ble adj.**

ORIGIN late Middle English: alteration of obsolete *scrow* 'roll.'

scroll bar /ˈskrōl ˌbär/ (also **scroll•bar**) ▸ **n.** a long thin section at the edge of a computer display by which material can be scrolled using a mouse.

scroll•er /ˈskrōlər/ ▸ **n.** a computer game in which the background scrolls past at a constant rate.

scroll•ing /ˈskrōliNG/ ▸ **n.** the action of moving displayed text or graphics up or down on a computer screen in order to view different parts of them.

SCSI /ˈskəzē/ ▸ **abbr.** small computer system interface, a bus standard for connecting computers and their peripherals together.

SDRAM /ˈes ˈdē ˌræm/ ▸ **abbr.** synchronous dynamic random-access memory, DRAM that synchronizes with the clock speed of the processor.

search en•gine /ˈsərCH ˌenjən/ ▸ **n.** a program for the retrieval of data, files, or documents from a database or network, esp. the Internet.

sec•tor /ˈsektər/ ▸ **n.** a subdivision of a track on a magnetic disk.

se•cure serv•er /siˈkyo͝or ˈsərvər/ ▸ **n.** an Internet server that encrypts confidential information supplied by visitors to Web pages.

se•cu•ri•ty patch /siˈkyo͝oritē ˌpæCH/ ▸ **n.** a software or operating-system patch that is intended to correct a vulnerability to hacking or viral infection: *Swen (also known as Gibe-F or Gibe-C) traveled during September attached to an e-mail purporting to be a security patch from Microsoft support.*

seek time /ˈsēk ˌtīm/ ▸ **n.** the time taken for a disk drive to locate the area on the disk where the data to be read is stored.

se•lect /səˈlekt/ ▸ **v.** [trans.] use a mouse or keystrokes to choose or mark (something) on a computer screen for a particular operation: *select the control panel from the menu, then click start.*

DERIVATIVES **se•lect•a•ble adj.**

ORIGIN mid 16th cent.: from Latin *select-* 'chosen,' from the verb *seligere*, from *se-* 'apart' + *legere* 'choose.'

se•lec•tion /sə'leksHən/ ▸ **n.** data highlighted on a computer screen that is a target for various manipulations: *the macro finds the text I told it to search for but it's not part of my original selection.*
■ the activity or capability of selecting data in this way.

se•quenc•er /'sēkwənsər/ ▸ **n.** a programmable electronic device for storing sequences of musical notes, chords, or rhythms and transmitting them when required to an electronic musical instrument.

se•quen•tial /si'kwenCHəl/ ▸ **adj.** performed or used in sequence: *sequential processing of data files.*
DERIVATIVES **se•quen•tial•ly adv.**

se•quen•tial ac•cess /si'kwenCHəl 'ækses/ ▸ **n.** access to a computer data file that requires the user to read through the file from the beginning in the order in which it is stored. Compare with **DIRECT ACCESS**.

se•ri•al /'si(ə)rēəl/ ▸ **adj.** (of a device) involving the transfer of data as a single sequence of bits. See also **SERIAL PORT**.
■ (of a processor) running only a single task, as opposed to multitasking.
DERIVATIVES **se•ri•al•ly adv.**
ORIGIN mid 19th cent.: from *series* + *-al*, perhaps suggested by French *sérial.*

se•ri•al port /'si(ə)rēəl 'pôrt/ ▸ **n.** a connector for a peripheral device that sends bits sequentially, one at a time. Compare **PARALLEL PORT**.

serv•er /'sərvər/ ▸ **n.** a computer or computer program that manages access to a centralized resource or service in a network.

serv•ice pack /'sərvis ˌpæk/ ▸ **n.** (abbreviation **SP**) a periodically released update to software from a manufacturer, consisting of requested enhancements and fixes to known bugs: *I had to update XP to Service Pack 1 for WebInspect to run.*

serv•ice pro•vid•er /'sərvis prəˌvīdər/ ▸ **n.** a company that provides its subscribers with access to the Internet: *a wireless service provider | a wireless service provider.*

serv•let /'sərvlit/ ▸ **n.** a small, server-resident program that typically runs automatically in response to user input: *students will learn to maximize Web application productivity by building servlets that generate Web pages, retrieve information, process data, communicate with applets, and communicate with other Java servers.*

set /set/ ▸ v. (**set•ting**; past and past part. **set**) [trans.] Electronics cause (a binary device) to enter the state representing the numeral 1.

ORIGIN Old English *settan*, of Germanic origin; related to Dutch *zetten*, German *setzen*.

SGML ▸ abbr. Standard Generalized Mark-up Language, an international standard for defining methods of encoding electronic texts to describe layout, structure, syntax, etc., which can then be used for analysis or to display the text in any desired format.

share•ware /ˈsHer,we(ə)r/ ▸ n. software that is available free of charge and often distributed informally for evaluation, after which a fee may be requested for continued use: *the $30 shareware is available as a 30-day demo.*

sheet feed•er /ˈsHēt ˌfēdər/ ▸ n. a device that feeds paper into a printer one sheet at a time.

shell /sHel/ (also **shell program**) ▸ n. a program that provides an interface between the user and the operating system: *the Windows shell.*

shell pro•gram /ˈsHel ˌprōgrəm; -græm/ ▸ n. a program that provides an interface between the user and the operating system.

shift /sHift/ ▸ v. [trans.] move (data) to the right or left in a register: *the partial remainder is shifted left.*

■ [intrans.] press the shift key on a keyboard.

▸ n. (also **shift key** /ˈshift ˌkē/) a key on a typewriter or computer keyboard used to switch between two sets of characters or functions, principally between lower- and upper-case letters.

■ a movement of the digits of a word in a register one or more places to left or right, equivalent to multiplying or dividing the corresponding number by a power of whatever number is the base.

short•cut /ˈsHôrt,kət/ ▸ n. an icon that represents a program or document, and when clicked, opens the program or document.

■ a combination of keys that select an item or carry out a command: *the application allows users to customize their keyboard shortcuts.*

shrink-wrap /ˈsHriNGk ˌræp/ ▸ v. [trans.] [as adjective] (**shrink-wrapped**) (of a product) sold commercially as a ready-made software package.

shut•down /ˈsHət,down/ ▸ **n.** a turning off of a computer or computer system.

SIG /sig/ ▸ **abbr.** special interest group, a type of newsgroup.

sig /sig/ ▸ **n.** informal an signature file, esp. one that contains a short message, quotation, etc., that can be inserted at the end of an e-mail message.
ORIGIN 1990s: abbreviation of **SIGNATURE.**

Sil•i•con Val•ley /ˈsili,kän ˈvælē/ a name given to an area between San Jose and Palo Alto in Santa Clara County, California, noted for its computing and electronics industries.

SIMM /sim/ ▸ **abbr.** single in-line memory module, containing RAM chips and having a 32-bit data path to the computer. Compare with **DIMM.**

sim•plex /ˈsim,pleks/ ▸ **adj.** (of a communications system, computer circuit, etc.) only allowing transmission of signals in one direction at a time.
ORIGIN late 16th cent.: from Latin, literally 'single,' variant of *simplus* 'simple.'

sim•u•late /ˈsimyə,lāt/ ▸ **v.** [trans.] produce a computer model of: *future population changes were simulated by computer.*
DERIVATIVES **sim•u•la•tion** /,simyəˈlāsHən/ **n.**
ORIGIN mid 17th cent.: from Latin *simulat-* 'copied, represented,' from the verb *simulare*, from *similis* 'like.'

sim•u•la•tor /ˈsimyə,lāṯər/ ▸ **n.** (also **sim•u•la•tor pro•gram**) a program enabling a computer to execute programs written for a different computer.

skin /skin/ ▸ **n.** a customized graphic user interface for an application or operating system: *the offerings here include music, reviews and attitude all wrapped up in the skin of a catalog.*

sleep /slēp/ ▸ **n.** turn off (devices attached to a computer) to reduce power use; hibernate or place on standby.
PHRASES **put to sleep** put a computer on standby while it is not being used.
ORIGIN Old English *slēp, slǣp* (noun), *slēpan, slǣpan* (verb), of Germanic origin; related to Dutch *slapen* and German *schlafen.*

slid•er /ˈslīdər/ ▸ **n.** an icon that can be dragged horizontally or vertically to control an adjustable element such as volume or brightness.

slide show /ˈslīd ˌSHō/ ▸ **n.** a series of text and graphics images created with presentation software and stored on a computer.

smart card /ˈsmärt ˌkärd/ ▸ **n.** a plastic card with a built-in microprocessor, used typically to perform financial transactions: *you can use your smart card at any store on campus.*

smart dust /ˈsmärt ˌdəst/ ▸ **n.** a collection of microelectromechanical systems forming a simple computer in a container light enough to remain suspended in air, used mainly for information gathering in environments that are hostile to life.

smart quotes /ˈsmärt ˌkwōts/ ▸ **plural n.** quotation marks that, although all keyed the same, are automatically interpreted and set as opening or closing marks rather than vertical lines.

smil•ey /ˈsmīlē/ ▸ **n.** a symbol that, when viewed sideways, represents a smiling face, formed by the characters :-) and used in electronic communications to indicate that the writer is pleased or joking; an emoticon.

SMTP ▸ **abbr.** simple mail transfer protocol, a data transmission format used to send and receive e-mail.

snail mail /ˈsnāl ˌmāl/ ▸ **n.** informal the ordinary postal system as opposed to e-mail.

■ correspondence sent using the postal system.

snap•shot /ˈsnæpˌSHät/ ▸ **n.** a record of the contents of a storage location or data file at a given time.

snert /snərt/ ▸ **n.** informal a participant in an Internet chat room who acts in a rude, annoying, or juvenile manner: *I could tell he was a snert from his sarcastic comments.* ■ a person whose online posts or e-mails are annoying to others: *do you ever get unsolicited messages from snerts?*

ORIGIN of uncertain origin, possibly an initialism from *snot-nosed egotistical rude twit* (or *teenager*).

sniff•er /ˈsnifər/ ▸ **n.** (also **sniff•er pro•gram**) a computer program that detects and records a variety of restricted information, esp. the secret passwords needed to gain access to files or networks.

SNOBOL /ˈsnōˌbôl/ ▸ **n.** a high-level computer programming language used esp. in manipulating textual data.

ORIGIN 1960s: formed from letters taken from *string-oriented symbolic language*, on the pattern of *COBOL.*

soft cop•y /ˈsôft ˈkäpē/ ▸ **n.** a legible version of a piece of information not printed on a physical medium, esp. as stored or displayed on a computer.

soft par•ti•tion•er /ˈsôft pärˈtisHənər/ ▸ see **PARTITIONER.**

soft•ware /ˈsôftˌwe(ə)r/ ▸ **n.** the programs and other operating information used by a computer. Compare with **HARDWARE.**

SOHO /ˈsōˌhō/ ▸ **adj.** relating to a market for relatively inexpensive consumer electronics used by individuals and small companies: *an all-in-one personal server designed to help consumers and SOHO professionals easily protect, remotely access and share their digital files.*

ORIGIN 1990s: acronym from *small office home office.*

sort /sôrt/ ▸ **n.** the arrangement of data in a prescribed sequence.

▸ **v.** [trans.] arrange (data) in a prescribed sequence.

DERIVATIVES **sort•a•ble adj.**; **sort•er n.**

ORIGIN late Middle English: from Old French *sorte*, from an alteration of Latin *sors, sort-* 'lot, condition.'

sor•ta•tion /sôrˈtāsHən/ ▸ **n.** (especially in data processing) the process of sorting or its result.

sound card /ˈsownd ˌkärd/ ▸ **n.** a printed circuit board controlling a computer's audio output.

source code /ˈsôrs ˌkōd/ ▸ **n.** a text listing of commands to be compiled or assembled into an executable computer program.

source pro•gram /ˈsôrs ˌprōgræm/ ▸ **n.** a program written in a language other than machine code, typically a high-level language.

SP ▸ **abbr.** service pack (usually followed by a number): *options within the code enable an attacker to specifically target either Windows 2000 SP3 and SP4.*

spam /spæm/ (**spammed**, **spam•ming**) ▸ **n.** unwanted, irrelevant, or inappropriate e-mail messages, usually sent to a large number of recipients.

▸ **v.** [trans.] send unwanted, irrelevant, or inappropriate e-mail messages indiscriminately to (recipients).

DERIVATIVES **spam•mer n.**

ORIGIN probably in reference to a 1971 sketch from *Monty*

Python's Flying Circus, set in a café where Spam was served as the main ingredient of every dish, and featuring a song whose lyrics were mainly the word "Spam" repeated over and over, interrupting or drowning out conversation.

spam fil•ter /ˈspæm ˌfiltər/ ▸ **n.** a program for detecting and intercepting spam before it is viewed or downloaded from an e-mail server to a user's e-mail inbox: *the spam filter will be updated monthly.*

■ a similar program that is part of an e-mail or anti-virus program: *a built-in spam filter.*

spawn /spôn/ ▸ **v.** [trans.] generate (a dependent or subordinate computer process): *from time to time it spawns two copies of the ip-up program, other times only one.*

speech rec•og•ni•tion /ˈspēCH ˌrekəgˌniSHən/ ▸ **n.** the process of enabling a computer to identify and respond to the sounds produced in human speech: [often as modifier] *a contact center automation solution using speech recognition technologies to streamline customer service.*

spell-check /ˈspel ˌCHek/ (also **spell check**) ▸ **v.** [trans.] check the spelling in (a text) using a spell-checker.

▸ **n.** a check of the spelling in a file of text using a spell-checker: *prerecorded macros for starting a spell-check or saving a file.*

■ a spell-checker.

spell-check•er /ˈspel ˌCHekər/ (also **spell check•er**) ▸ **n.** a computer program that checks the spelling of words in files of text, typically by comparison with a stored list of words.

spell•ing check•er /ˈspeliNG ˌCHekər/ ▸ **n.** another term for **SPELL-CHECKER.**

spi•der /ˈspīdər/ ▸ **n.** a program that searches and indexes the World Wide Web.

spin-down /ˈspin ˌdown/ ▸ **n.** a decrease in the speed of rotation of a spinning object, in particular a heavenly body or a computer disk.

split screen /ˈsplit ˈskrēn/ ▸ **n.** a computer screen on which two or more separate windows are displayed.

spool /spo͞ol/ ▸ **v.** [trans.] send (data that is intended for printing or processing on a peripheral device) to an intermediate store: *users can set which folder they wish to spool files to.*

ORIGIN acronym from *simultaneous peripheral operation online.*

spread•sheet /ˈspred,SHēt/ ▸ **n.** a computer program used chiefly for accounting, in which figures arranged in the rows and columns of a grid can be manipulated and used in calculations.

▸ **v.** [intrans.] [usu. as noun] (**spread•sheet•ing**) use such a computer program.

sprite /sprīt/ ▸ **n.** a computer graphic that may be moved on-screen and otherwise manipulated as a single entity: *animated sprites.*

SQL ▸ **abbr.** Structured Query Language, a standard for database manipulation.

SRAM /ˈes ˌræm/ ▸ **abbr.** static random-access memory.

SSL ▸ **abbr.** Secure Sockets Layer, a computing protocol that ensures the security of data sent via the Internet by using encryption.

stack /stæk/ ▸ **n.** a set of storage locations that store data in such a way that the most recently stored item is the first to be retrieved.

ORIGIN Middle English: from Old Norse *stakkr* 'haystack,' of Germanic origin.

stacked /stækt/ ▸ **adj.** (of a task) placed in a queue for subsequent processing.

■ (of a stream of data) stored in such a way that the most recently stored item is the first to be retrieved.

stand-a•lone /ˈstænd əˌlōn/ (also **stand•a•lone**) ▸ **adj.** (of computer hardware or software) able to operate independently of other hardware or software.

stand•by /ˈstændˌbī/ ▸ **n.** a condition in which a computer is ready for immediate use, but its monitor and hard drives have automatically turned off to reduce power use. Compare with **HIBERNATE.**

star /stär/ ▸ **n.** (also **star net•work**) [usu. as adjective] a data or communication network in which each device is independently connected to one central unit: *computers in a star layout.*

ORIGIN Old English *steorra*, of Germanic origin; related to Dutch *ster*, German *Stern*, from an Indo-European root shared by Latin *stella* and Greek *astēr*.

stat•ic /ˈstæt̲ik/ ▸ **adj. 1** (of a memory device) not needing to be periodically refreshed by an applied voltage.

2 (of a process or variable) not able to be changed during a set period, for example while a program is running.

ORIGIN late 16th cent. (denoting the science of weight and its effects): via modern Latin from Greek *statikē (tekhnē)* 'science of weighing'; the adjective from modern Latin *staticus*, from Greek *statikos* 'causing to stand,' from the verb *histanai*.

sta•tus bar /ˈstætəs ˌbär; ˈstātəs/ ▸ **n.** a horizontal bar, typically at the bottom of the screen or window, showing information about a document being edited or the currently active programs.

stick•y /ˈstikē/ ▸ **adj.** (**stick•i•er**, **stick•i•est**) informal (of a Web site) attracting a long visit or repeat visits from users: *experts measure the attractiveness of pages by how sticky they are.*

stor•age /ˈstôrij/ ▸ **n.** the retention of retrievable data on a computer or other electronic system; memory.

stor•age de•vice /ˈstôrij diˌvīs/ ▸ **n.** a piece of computer equipment on which information can be stored.

store /stôr/ ▸ **v.** [trans.] retain or enter (information) for future electronic retrieval: *the data is stored on disk.*

ORIGIN Middle English: shortening of Old French *estore* (noun), *estorer* (verb), from Latin *instaurare* 'renew'.

stream /strēm/ ▸ **n.** a continuous flow of data or instructions, typically one having a constant or predictable rate.

ORIGIN Old English *strēam* (noun), of Germanic origin; related to Dutch *stroom*, German *Strom*, from an Indo-European root shared by Greek *rhein* 'to flow.'

stream•ing /ˈstrēmiNG/ ▸ **n.** a method of relaying data (especially video and audio material) over a computer network as a steady continuous stream: *streaming means that the first part of the audio file is saved in memory, then played while the rest of the soundfile is simultaneously downloading.*

▸**adj.** (of data) transmitted in a continuous stream: *adding streaming audio, video, or data to a site is not brain surgery.*

strike•out /ˈstrīkˌowt/ ▸ **adj.** Computing (of text) having a horizontal line through the middle; crossed out.

string /striNG/ ▸ **n.** a linear sequence of characters, words, or other data.

style sheet /ˈstīl ˌSHēt/ ▸ **n.** a file consisting of settings font, layout, or other details to give a standardized look to certain documents.

sty•lus /ˈstīləs/ ▸ **n.** (pl. **sty•li** /ˈstīlī/ or **sty•lus•es**) a penlike device used to input handwritten text or drawings directly into a computer or for input on a touch-senstitive display screen: *this handheld comes with a spare stylus.*

ORIGIN early 18th cent. erroneous spelling of Latin *stilus.*

sub•di•rec•to•ry /ˈsəbdiˌrektərē/ ▸ **n.** (pl. **sub•di•rec•to•ries**) a directory below another directory in a hierarchy.

sub•men•u /ˈsəbˌmenyo͞o/ ▸ **n.** a menu of specific items accessed from a more general menu: *there is a submenu for CD recording.*

sub•net•work /səbˈnetˌwərk/ (also **sub•net** /səbˈnet/) ▸ **n.** a part of a larger network such as the Internet.

sub•pro•gram /ˈsəbˌprōgræm/ ▸ **n.** another term for **SUBROUTINE**.

sub•rou•tine /ˈsəbro͞oˌtēn/ ▸ **n.** a set of instructions designed to perform a frequently used operation within a program.

sub•script /ˈsəbˌskript/ ▸ **adj.** (of a letter, figure, or symbol) written or printed below the line.

▸ **n.** a symbol (notionally written as a subscript but in practice usually not) used in a program, alone or with others, to specify one of the elements of an array.

ORIGIN early 18th cent.: from Latin *subscript-* 'written below,' from the verb *subscribere.*

sub•web /ˈsəbˌweb/ ▸ **n.** an isolated part of a Web site, especially one that is password protected or that is not obviously accessible from the main page: *the hit counter will not work in a subweb on a Netscape Enterprise NT.*

suite /swēt/ ▸ **n.** a set of programs with a uniform design and the ability to share data.

ORIGIN late 17th cent.: from French, from Anglo-Norman French *siwte.*

su•per•com•put•er /ˈso͞opərkəmˌpyo͞otər/ ▸ **n.** the fastest and most powerful type of computer, used in science and engineering: *the military is to buy a new supercomputer | a parallel processing supercomputer.*

DERIVATIVES **su•per•com•put•ing n.**

su•per•high•way /ˈso͞opərˌhīwā; ˌso͞opərˈhīˌwā/ ▸ **n.** see **INFORMATION SUPERHIGHWAY**.

su•per•sca•lar /ˌso͞opərˈskālər/ ▸ **adj.** denoting a computer architecture where several instructions are loaded at once and, as far as possible, are executed simultaneously, shortening the time taken to run the whole program: *32-bit RISC superscalar architecture.*

sup•port /sə'pôrt/ ▸ **v.** [trans.] (of a computer or operating system) allow the use or operation of (a program, language, or device): *the new versions do not support the graphical user interface standard.*
▸ **n.** technical help given to the user of a computer or other product.
DERIVATIVES **sup•port•a•bil•i•ty n.; sup•port•a•ble adj.**

surf /sərf/ ▸ **v.** move from site to site on (the Internet).
DERIVATIVES **surf•er n.**
ORIGIN late 17th cent.: apparently from obsolete *suff*, of unknown origin, perhaps influenced by the spelling of *surge*.

SVGA ▸ **abbr.** super video graphics array, a high-resolution standard for monitors and screens.

swap•file /'swäp,fīl/ ▸ **n.** a file on a hard disk used to provide additional memory space for programs that have been transferred out of active memory.

switch /swiCH/ ▸ **n.** a program variable that activates or deactivates a certain function of a program.
DERIVATIVES **switch•a•ble adj.**
ORIGIN late 16th cent. (denoting a thin tapering riding whip): probably from Low German.

SXGA ▸ **abbr.** super extended graphics array, a standard very for high-resolution monitors and screens.

syn•tax er•ror /'sin,taks ,erər/ ▸ **n.** a character or string incorrectly placed in a command or instruction that causes a failure in execution: *I am trying to create a List implementation using templates, and am getting a syntax error.*

syn•thes•pi•an /sin'THespēən/ ▸ **n.** a computer-generated actor appearing in a film with human actors and interacting with them or in a wholly animated film.

sys•op /'sis,äp/ ▸ **n.** a system operator.
ORIGIN 1980s: abbreviation.

sys•tem /'sistəm/ ▸ **n.** a group of related software programs sold together: *an integrated personal information management suite.*

ORIGIN early 17th cent.: from French *système* or late Latin *systema*, from Greek *sustēma*, from *sun-* 'with' + *histanai* 'set up.'

sys•tem op•er•a•tor /ˈsistəm ˌäpərātər/ (also **sys•tems op•er•a•tor**)
▸ **n.** a person who manages the operation of a computer system, such as an electronic bulletin board.

T

T-1 ▸ **n.** a high-speed data line capable of transmitting at approximately 1.5 million bps: [as modifier] *on an intranet where everyone is connected at T1 rates or higher, you can achieve high video quality in quarter-screen windows.*

tab /tæb/ ▸ **n.** a facility in a word-processing program, or a device on a typewriter, used for advancing to a sequence of set positions in tabular work: *set tabs at 1.4 inches and 3.4 inches.*

▸ v. (**tabbed**, **tab•bing**) activate the tab feature on a word processor or typewriter: *the user can tab to the phrase and press Enter.*

ta•ble /ˈtābəl/ ▸ **n.** a collection of data stored in memory as a series of records, each defined by a unique key stored with it.

ORIGIN Old English *tabule* 'flat slab, inscribed tablet,' from Latin *tabula* 'plank, tablet, list,' reinforced in Middle English by Old French *table.*

tab•let /ˈtæblit/ (also **tablet PC** /ˈtæblit ˈpē ˈsē/) ▸ **n.** a small battery-powered portable computer with handwriting recognition software and a screen on which a user can write notes using a digital pen or stylus: *the school is selling each incoming freshman a wireless tablet PC that they will use in classes to take notes.*

tab•let PC /ˈtablit ˈpēˈsē/ ▸ **n.** a microcomputer that accepts input directly onto an LCD screen by means of a stylus, savable as image or text: *ever since Bill Gates started waving around a prototype of Microsoft's Tablet PC, consumers have been lusting after it: a notepad-sized digital slate that responds to voice commands, takes notes, is always wirelessly connected to the Internet, and can do anything a desktop computer can.*

tag /tæg/ ▸ **n.** a character or set of characters appended to an item of data in order to identify it.

▸ **v.** (**tagged**, **tag•ging**) [trans.] add a character or set of characters to (an item of data) in order to identify it for later retrieval.

TB (also **Tb**) ▸ **abbr.** terabyte(s).

Tcl /ˈtikəl/ ▸ **n.** a computer programming language used to connect other languages.

ORIGIN 1990: acronym for Tool Control Language.

TCP/IP ▸ **abbr.** transmission control protocol/Internet protocol, a method of transmitting data over the Internet which divides messages into packets of two layers, one containing a portion of the message and the other the address to which it is being sent, where the message is reassembled.

teach•ing ma•chine /ˈtēCHiNG məˌSHēn/ ▸ **n.** a machine or computer that gives instruction to a student according to a program, reacting to their responses.

tech•ni•cal sup•port /ˈteknikəl səˈpôrt/ ▸ **n.** a service provided by a hardware or software company that provides registered users with help and advice about their products: *24-hour technical support* | *technical support for corporate customers.*

■ a department within an organization that maintains and repairs computers and computer networks.

tech•nol•o•gy trans•fer /tekˈnäləjē ˌtrænsfər/ ▸ **n.** the transfer of new technology from the originator to a user, as from research centers to private industry or from developed to less developed countries: *the public benefit and economic impact of technology transfer.*

tel•e•con•fer•ence /ˈteliˌkänf(ə)rəns/ ▸ **n.** a conference with participants in different locations linked by telecommunications devices.

▸ **v.** [intrans.] participate in a teleconference: *he teleconferenced with everyone who had been in attendance.*

DERIVATIVES **tel•e•con•fer•enc•ing n.**

tel•e-im•mer•sion /ˌtelə iˈmərZHən/ ▸ **n.** two-way remote communication in which each party gets an audio and three-dimensional visual representation of the other, via high-speed data exchange: *tele-immersion allows users to climb into a computer screen.*

tel•e•port /ˈteləˌpôrt/ ▸ **n.** a center providing interconnections between different forms of telecommunications, esp. one that links satellites to ground-based communications: *the company their satellite networking service at a teleport in Virginia.*

ORIGIN 1980s: originally the name of such a center in New York, from (1950s) back-formation from *teleportation* (1930s), from *tele-* 'at a distance' + a shortened form of *transportation.*

tel•e•pres•ence /ˈteləˌprezəns/ ▸ **n.** the use of virtual reality technology, esp. for remote control of machinery or for apparent participation in distant events: *telepresence surgery enables a surgeon to give advice from a remote location on an operation taking place in the mobile unit.*

tel•net /ˈtelˌnet/ ▸ **n.** a network protocol that allows a user on one computer to log on to another computer that is part of the same network.

■ a program that establishes a connection from one computer to another by means of such a protocol: *access your e-mail using telnet.* ■ a link established in such a way.

▸ **v.** (**tel•net•ted, tel•net•ting**) [intrans.] informal log on to a remote computer using a telnet program: *you'll be able to telnet the server.*

DERIVATIVES **tel•net•ta•ble adj.**

ORIGIN 1970s: blend of *telecommunication* and **NETWORK.**

tera- /ˈterə/ ▸ **comb. form** used in units of measurement: **1** denoting a factor of 2^{40}.

2 denoting a factor of 10^{12}: *terawatt.*

ORIGIN from Greek *teras* 'monster.'

ter•a•byte /ˈterəˌbīt/ (abbr.: **Tb** or **TB**) ▸ **n.** a unit of information equal to one million million (10^{12}) or strictly, 2^{40} bytes.

ter•a•flop /ˈterəˌfläp/ ▸ **n.** a unit of computing speed equal to one million million floating-point operations per second.

ter•mi•nal /ˈtərmənl/ ▸ **n.** a device connected to a network computer at which a user enters data or commands and that displays the received output.

ORIGIN early 19th cent.: from Latin *terminalis*, from *terminus* 'end, boundary.'

text /tekst/ ▸ **n.** data in written form, esp. when stored, processed, or displayed in a word processor.

ORIGIN late Middle English: from Old Northern French *texte*, from Latin *textus* 'tissue, literary style' (in medieval Latin, 'Gospel'), from *text-* 'woven,' from the verb *texere*.

text ed•i•tor /ˈtekst ˌeditər/ ▸ **n.** a system or program that allows a user to edit text, often part of a word processing package: *if you are writing your own HTML code, you'll need a text editor.*

text proc•ess•ing /ˈtekst ˌpräsesiNG/ ▸ **n.** the manipulation of text, esp. the transformation of text from one format to another.

tex•ture map•ping /ˈteksCHər ˌmæpiNG / ▸ **n.** the application of patterns or images to three-dimensional graphics to enhance the realism of their surfaces.

text wrap /ˈtekst ˌræp/ ▸ **n.** a software facility allowing text to surround embedded features such as pictures.

TFT ▸ **abbr.** thin-film transistor, an active-matrix technology for high-resolution LCD color screens in which transistors control each pixel, used in flat-panel displays and portable computers.

thread /THred/ ▸ **n.** a group of linked messages posted on an Internet discussion area, such as a message board, that share a common subject or theme: *a discussion thread on the Classic Movies Forum.*

■ a programming structure or process formed by linking a number of separate elements or subroutines, esp. each of the tasks executed concurrently in multithreading.

DERIVATIVES **thread•like adj.**

ORIGIN Old English *thrǣd* (noun), of Germanic origin; related to Dutch *draad* and German *Draht*.

through•put /ˈTHro͞oˌpo͝ot/ ▸ **n.** the amount of material or items passing through a system or process.

thumb•nail /ˈTHəmˌnāl/ ▸ **n.** a small picture of an image or page layout.

TIFF /tif/ ▸ **abbr.** tagged image file format, a format for storing images that is widely used in desktop publishing.

tile /tīl/ ▸ **v.** [trans.] (usu. **be tiled**) arrange (two or more windows) on a computer screen so that they do not overlap.

ORIGIN Old English *tigele*, from Latin *tegula*, from an Indo-European root meaning 'cover.'

til•ing /ˈtīliNG/ ▸ **n.** a technique for displaying several nonoverlapping windows on a computer screen.

time /tīm/ ▸ **v.** [trans.] (**time something out**) (of a computer or a program) cancel an operation automatically because a predefined interval of time has passed without a certain event happening.

time lock /ˈtīm ˌläk/ ▸ **n.** a device built into a computer program to stop it operating after a certain time.

time•out /ˈtīmˈowt/ (also **time-out**) ▸ **n.** an automatic cancellation of further activity by a program when it hasn't received input for a set period of time: [as adjective] *a timeout period.*

time-shar•ing /ˈtīm ˌSHe(ə)riNG/ ▸ **n.** the operation of a computer system by several users for different operations at the same time.

ti•tle bar /ˈtītl ˌbär/ ▸ **n.** a horizontal bar at the top of a window, bearing the name of the program and typically the name of the currently active document.

tog•gle /ˈtägəl/ ▸ **n.** (also **tog•gle switch** or **tog•gle key**) a key or command that causes the computer or program to switch to one of two features, views, effects, etc., or back again.

▸ **v.** [intrans.] switch from one effect, feature, or state to another by using a toggle: *you should be able to toggle back and forth between the English and Swedish keyboards.*

ORIGIN mid 18th cent. (originally in nautical use): of unknown origin.

tog•gle switch /ˈtägəl ˌswiCH/ ▸ **n.** **1** another term for **TOGGLE**.

2 an electric switch operated by means of a projecting lever that is moved up and down.

to•ken /ˈtōkən/ ▸ **n.** **1** the smallest meaningful unit of information in a sequence of data for a compiler.

2 a sequence of bits used in some network architecture in which the ability to transmit information is conferred on a particular node by the arrival there of this sequence, which is passed continuously between nodes in a fixed order.

ORIGIN Old English *tāc(e)n*, of Germanic origin; related to Dutch *teken* and German *Zeichen.*

to•ken ring /ˈtōkən ˌriNG/ ▸ **n.** a local area network in which a node can transmit only when in possession of a sequence of bits (called the token) that is passed to each node in turn.

tool /to͞ol/ ▸ **n.** a piece of software that carries out a particular function, typically creating or modifying another program.

ORIGIN Old English *tōl*, from a Germanic base meaning 'prepare'.

tool•bar /'to͞ol,bär/ ▸ **n.** a strip of icons in a program used to perform certain functions: *the toolbar of this program features some customizable options.*

tool•box /'to͞ol,bäks/ ▸ **n.** a set of software tools.

■ the set of programs or functions accessible from a single menu.

tool kit /'to͞ol ,kit/ ▸ **n.** a set of software tools.

touch•pad /'təCH,pæd/ ▸ **n.** a small touch-sensitive panel on a notebook computer that you drag a finger across to move the cursor on the screen.

touch•point /'təCH,point/ ▸ **n.** a device like a miniature joystick with a rubber tip, manipulated with a finger to move the screen pointer on some laptop computers.

touch screen /'təCH ,skrēn/ (also **touch•screen** /'təCH,skrēn/) ▸ **n.** a display device that allows a user to interact with a computer by touching areas on the screen.

tow•er /'tow-ər/ ▸ **n.** a vertical computer case.

trace /trās/ ▸ **n. 1** a procedure to investigate the origin of an error in a computer program.

2 a line or pattern displayed by an instrument using a moving pen or a luminous spot on a screen to show the existence or nature of something that is being investigated.

DERIVATIVES **trace•a•bil•i•ty** /,trāsə'bilitē/ **n.**; **trace•a•ble adj.**; **trace•less adj.**

ORIGIN Middle English (first recorded as a noun in the sense 'path that someone or something takes'): from Old French *trace* (noun), *tracier* (verb), based on Latin *tractus*.

track•ball /'træk,bôl/ ▸ **n.** a small ball set in a holder that can be rotated by hand to move a cursor on a computer screen.

trans•ac•tion /træn'sækSHən/ ▸ **n.** an input message to a computer system that must be dealt with as a single unit of work.

ORIGIN late Middle English (as a term in Roman law): from late Latin *transactio(n-)*, from *transigere* 'drive through'.

trans•ceiv•er /træn'sēvər/ ▸ **n.** a device that can both transmit and receive communications, in particular a combined radio transmitter and receiver.

ORIGIN 1930s: blend of *transmitter* and *receiver*.

trans•la•tor /'træns,lāt̲ər/ ▸ **n.** a program that translates from one programming language into another.

trans•par•ent /træn'spe(ə)rənt/ ▸ **adj.** (of a process or interface) functioning without the user being aware of its presence.

DERIVATIVES **trans•par•ent•ly adv.** [as submodifier] *a transparently feeble argument.*

ORIGIN late Middle English: from Old French, from medieval Latin *transparent-* 'shining through,' from Latin *transparere*, from *trans-* 'through' + *parere* 'appear.'

trap•door /'træp'dôr/ (also **trap door**) ▸ **n.** a feature or defect of a computer system that allows surreptitious unauthorized access to data belonging to other users.

trash /træsʜ/ ▸ **v.** [trans.] kill (a file or process) or wipe (a disk): *she almost trashed the e-mail window.*

ORIGIN late Middle English: of unknown origin.

tree struc•ture /'trē ,strəkcʜər/ ▸ **n.** a structure that has successive branchings or subdivisions.

Tro•jan Horse /'trōjən 'hôrs/ (also **Tro•jan horse** or **Tro•jan**) ▸ **n.** a program designed to breach the security of a computer system while ostensibly performing some innocuous function. *a Trojan horse hidden in an e-mail*

troll /trōl/ informal ▸ **v.** [intrans.] send (an e-mail message or posting on the Internet) intended to provoke a response from the reader by containing errors.

▸ **n.** an e-mail message or posting on the Internet intended to provoke a response in the reader by containing errors.

DERIVATIVES **troll•er n.**

ORIGIN late Middle English : origin uncertain; compare with Old French *troller* 'wander here and there (in search of game)' and Middle High German *trollen* 'stroll.'

TSR ▸ **abbr.** terminate and stay resident, denoting a type of program that remains in the memory of a computer after it has finished running and which can be quickly reactivated.

TTS ▸ **abbr.** text-to-speech, a form of speech synthesis used to create a spoken version of the text in an electronic document: *the agency then uses TTS to turn the text into audio within minutes, saving the company both time and money.*

Tu•ring ma•chine /ˈt(y)o͝oriNG məˌSHēn/ ▸ **n.** a mathematical model of a hypothetical computing machine that can use a predefined set of rules to determine a result from a set of input variables.

Tu•ring test /ˈt(y)o͝oriNG ˌtest/ ▸ **n.** a test of artificial intelligence devices or programs in which the goal is to make a human interacting electronically unable to tell whether it is a machine or another human: *the Turing test was held here at the behest of New York businessman Hugh Loebner, president of restaurant supplier Crown Industries Inc., who has offered a $100,000 prize for the first computer system able to pass it.*

ORIGIN after mathematician Alan *Turing*, (1912–1954), who proposed it in his 1950 paper, 'Computing Machinery and Intelligence.'

twist•ed pair /ˈtwistid ˈpe(ə)r/ ▸ **n.** a cable consisting of two wires twisted around each other, used especially for telephone or computer applications: *we prewired the building with eight-strand twisted pair with half-floor granularity.*

THE TEN BEST TOOLS AND PERIPHERALS YOU DIDN'T KNOW ABOUT (OR DIDN'T KNOW YOU WANTED)

1. **Flat-panel monitors**—OK, you not only know about flat-panel monitors, you've longed for one for years. But if you haven't checked the prices recently, check again—this may be the time to buy. These monitors are now brighter, sharper, and clearer than ever. A 17-, 19-, even a 20-inch display has become more affordable, especially as part of a new computer system. With no border inside the frame, you see much more display area than with a CRT that boasts the same screen size. Some can even be rotated between portrait and landscape views. In the bargain, *flat-panel monitors* weigh less, emit fewer UV rays, use less electricity, and generate far less heat. They really *are* cool!

2. **Multifunction printers**—*MFPs (multi-function printers)*, or *all-in-ones,* actually *do* do it all: print, copy, scan, and (usually) fax. Conventional wisdom cautions against hardware that performs more than one function, on the theory that if one part fails the whole thing goes south. But modern MFPs are solid and reliable. And they're vastly more simple to set up and use than four separate machines (think software, wires, and space). Laser MFPs usually print, copy, and fax in black and white but

scan in color. Ink-jet models do everything in color. Ink jets are initially less expensive, but if you factor in the cost of rapidly consumed color cartridges, laser is cheaper in the long run—especially if your primary output is text.

3. **Photo printers**—On the other hand, specialized ink-jet photo printers, which make it a breeze to print color images from your digital camera, are growing more and more popular. Some of them are flexible enough to print *directly*—not only from digital cameras but from memory cards and wireless devices, like PDAs and camera phones. Others connect through your PC. Some even allow you to print labels and photos onto special CDs! But primarily, you are freed to print stunningly clear, professional-looking, borderless pictures without leaving home. See brands from, among others, Epson, Canon, and Hewlett Packard.

4. **Keyboards for everyone**—For those who can't abide the flimsy $4.00 keyboards that come with most computers, third-party keyboards abound—wireless; Bluetooth wireless; multimedia; ergonomically sculpted; split; folded (for PDAs), one-handed; programmable; with function keys on the side (remember those?); with tactile, clicky keys; or with a quirky (not QWERTY) key layout (look for Dvorak or PLUM keyboards on the Internet). For more choices than you'll believe, see http://www.ergo-2000.com/. While you're there, check out the remarkable Comfort Keyboard System. It's split into three connected sections, allowing people who are wide-shouldered to stretch the keyboard sideways. Each section tilts, rotates, lifts, or lowers independently. The Kinesis Maxim Adjustable Ergonomic Keyboard, in two sections, is almost as flexible. Who could ask for more?

5. **Mice**—In recent years, three innovations have made the lowly rodent a peripheral to covet. First came *the scroll wheel,* usually placed between the right and left buttons on a two-

button mouse. It not only lets you scroll easily through a document, it can adjust the size of text in your word processor, in incoming e-mail, or on some Web sites (hold down the Ctrl key and rotate the wheel). Next was *the optical mouse,* which has no rolling mechanical parts and thereby eliminates the persistent lint collection under your mouse ball. Get one that fits your hand and feels responsive. Then came *the wireless mouse,* which controls your pointer through radio waves and effortlessly resolves those too-short-cord problems. It is also not trivial that you can switch hands easily if that helps you to stave off carpal tunnel syndrome. Look for rechargeable batteries. All of these mice are nice, and many come with extra programmable buttons. But a combination—*a cordless optical mouse with scroll wheel*—is hard to beat. They come in various sizes, including more than one minimouse for laptops. Logitech makes several models, and Gyration's mouse, with its "gyroscopic motion sensor," allows you to alternate between desktop and midair use! No floating mousepad required.

6. **Videoconferencing cameras**—Videoconferencing and video instant messaging, which used to be the stuff of science fiction, have been around for a while. But this is not your father's camera. *The Logitech QuickCam Orbit* doesn't just sit quietly on your desktop or monitor. It watches you and tracks your face as you move! Reach for your briefcase, take notes on your PDA, stretch, and you still remain in the center of the video window—so long as you don't go out of range. (The camera can swivel 128 degrees from side to side and 54 degrees up and down.) According to Logitech, the accompanying software is simple—with no "remote monitoring, surveillance, or video-editing." It simply frees you to be yourself while communicating from afar, using MSN Messenger, Windows Messenger, Yahoo!, or an included video e-mail client.

7. **DVD burners**—If you don't have a DVD burner (or writer) yet, you soon will. Prices no longer hover at more than $500, and the "plus" and "dash" format confusion (DVD-R vs.

DVD+R and DVD-RW vs. DVD+RW) has been resolved with the emergence of burners that support more than one available format. More good news is the emergence of plug-and-play *external* drives as well as *stand-alone* or *set-top* recorders that work with your TV. Internal or External, USB or FireWire, DVD burners reflect their revised name; they are indeed digital *versatile* drives—able to capture video not just from your PC but also from your camcorder or your old VHS tapes. There is software to edit data, music, video, and photos, and the drives read and write to CDs. Vendors include LaCie, Plextor, Memorex, Pioneer, HP, Panasonic, Philips, and Sony. The Next Generation? It's coming. High definition and capacities of up to 30 GB per disk.

8. **External hard drives**—Have you become cavalier about backups? Although making them can seem tedious, what with a gigantic hard drive in your PC—filled with bloated applications and irreplaceable data—y o u s h o u l d b a c k u p! And now you can forget multiple CDs; toss the floppies; throw your old, inadequate tape drive away; and get an easy-to-use external hard drive that allows you to back up not only your data files but your entire multi-gig system. Happily, these sleek, well-designed drives, with fast data transfer rates, have suddenly proliferated. They take up little space but typically come in capacities ranging from 80GB to 250GB. Connecting to your PC or Mac through FireWire or USB ports, they come complete with backup software. Don't just back up your My Documents folder; your data files include music, pictures, video, e-mails, and your browser favorites (or bookmarks). Available brands include the usual suspects—Seagate, Maxtor, Simpletech, Western Digital, and LaCie (designed by Porche!).

9. **USB flash drives**—The long awaited demise of the floppy has been made inevitable by these appealing, convenient, thumb-sized, highly portable removable hard drives. Plug one into any desktop or notebook that has a USB port and the drive is magically available in your Explorer window for file

downloading or uploading. No drivers are needed with Windows ME, 2000, or XP; drivers for Win 98 and 98 SE are supplied. These drives work equally well on a Windows or Linux machine or a Mac and come with capacities from 16 to 512MB. They're fast and versatile, with seemingly unlimited functions in their future. Software already exists to partition the disk and encrypt part of it for secure file transfer. There is at least one self-contained e-mail application available (nPOPq) that allows you to check e-mail accounts without installing anything on the computer you've plugged into, and more applications are promised. There are only two negatives: (1) their size makes them easy to lose (try to get one that's brightly colored or that slips onto a keyring) and (2) they're not yet bootable in the event of a system crash. (Maybe you *do* still need a floppy.) Brands include Belkin, Cables to Go, Fuji, Iomega, Kingston, Linksys, SimpleTech, Sony, and M-Systems, which makes the DiskOnKey.

10. **Software utilities**—Software, too, is a tool, and there are surprisingly good utilities out there—freeware, shareware, and inexpensive payware—that can make you fall to your knees in gratitude. (You already know about the famous ones, like *Adobe Reader, WinZip, and ZoneAlarm*). A few examples of some less well-known apps: (1) *InCtrl5,* a powerful utility available from http://www.pcmag.com/article2/0,4149,25126,00.asp, puts you in control of your computer system by letting you monitor and record installations, uninstalls, and any other changes to your system that might possibly cause trouble. You have to join *PC Magazine*'s utility library (for as little $5 for three downloads) in order to download the program, but it's well worth it. *InCtrl5's* reports make real troubleshooting possible by telling you exactly which files and registry entries were added, deleted, or changed by any procedure you choose to monitor. What went wrong? What did that last program do to make something else stop working? Can you undo the damage? *InCtrl5* gives you answers. Those who have it consider it a lifesaver. (2) Are you

tired of endlessly repeating the same keystrokes or drilling down the same menu tree over and over? For about $40 from http://www.macros.com, *Macro Express* lets you easily create, record, edit, and play mouse and keyboard macros across or within programs, including your browser. A single hotkey or shortcut can take you to an Internet site and log you in. Just think—you'll never have to type your credit card number at a new shopping site again! (3) How do you know if your backup copy of critical files got them right? *DrinkCompare* is an amazingly quick, sweet little program that compares two files or two entire directory trees and highlights any files that are not exactly the same. And it's free. Get it at http://digilander.libero.it/drinka/. Click at the upper left corner of the screen for the English version. (4) Got a new computer? Can you face the nightmare of transferring your old stuff to the new machine? Detto Technologies' *IntelliMover* (a $40 download at http://www.detto.com; slightly more for the boxed software with cable) is a data migration tool that "lets you quickly and easily move your e-mail, music, photos, files, folders, settings, and more from an old computer to a new one." Unlike some programs, this one lets you control what's moved. And if you forgot something, just go back and fetch it. Detto's similar *Move2Mac* provides the same service to Apple users, including moving files and settings from a PC. Note: for *any* utility, check for compatibility with your operating system and always make sure you are dealing with a reliable source.

U

un•bun•dle /ˌənˈbəndl/ ▸**v.** [trans.] market or charge for (software, hardware, or services) separately rather than as part of a package.
DERIVATIVES **un•bun•dler n.**

un•com•ment /ˌənˈkäˌment/ ▸**v.** [trans.] change (a piece of text within a program) from being a comment to being part of the program that is run by the computer.

un•de•lete /ˌəndiˈlēt/ ▸**v.** [trans.] cancel the deletion of (text or a file).

un•der /ˈəndər/ ▸**prep.** within the environment of (a particular operating system): *the program runs under UNIX.*
ORIGIN Old English, of Germanic origin; related to Dutch *onder* and German *unter.*

un•do /ˌənˈdo͞o/ ▸**v.** (**un•does** /ˌənˈdəz/; past **un•did** /ˌənˈdid/; past part. **un•done** /ˌənˈdən/) [trans.] cancel (the last command executed by a computer).
▸**n.** a feature of a computer program that allows a user to cancel or reverse the last command executed.

un•group /ˌənˈgro͞op/ ▸**v.** [trans.] separate (items) from a group formed within a word-processing or graphics package.

u•ni•cast /ˈyo͞oniˌkæst/ ▸**n.** transmission of a data package or an audio/visual signal to a single recipient: [as modifier] *the unicast method wastes a lot of bandwidth by sending duplicate information.*
ORIGIN 1990s: on the pattern of *broadcast.*

un•in•stall /ˌəninˈstôl/ ▸**v.** [trans.] remove (an application or file) from a computer: *if you wanted to uninstall them, you could never be certain which files could be safely deleted* | [intrans.] *the main drawback to DoubleSpace is that it is almost impossible to uninstall.*
▸**adj.** denoting a command, function, or capability to remove soft-

ware: [as modifier] *at a minimum, this means that you use the registry, not add information to WIN.INI or SYSTEM.INI, and provide complete uninstall capability with your application.*

DERIVATIVES **un•in•stall•er n.**

un•in•ter•rupt•i•ble pow•er sup•ply /ˈənintəˈrəptəbəl ˈpow-ər səˌplī/ (abbr.: **UPS**) ▸ **n.** a battery backup device that powers a computer for a short time in the event of a power failure.

UNIX /ˈyo͞oniks/ (also **Unix**) ▸ **trademark** a widely used multiuser operating system.

ORIGIN 1970s: from *uni-* 'one' + a respelling of *-ics*, on the pattern of an earlier less compact system called *Multics*.

un•pack /ˌənˈpæk/ ▸ **v.** [trans.] convert (data) from a compressed form to a usable form.

un•pro•tect•ed /ˌənprəˈtektid/ ▸ **adj.** (of data or a memory location) able to be accessed or used without restriction.

un•pruned /ˌənˈpro͞ond/ ▸ **adj.** not subjected to any reducing, trimming, or refining process: *BA work cycles aim to present structured, unpruned data to the domain experts.*

un•sub•scribe /ˌənsəbˈskrīb/ ▸ **verb** [intrans.] cancel a subscription to an Internet newsletter, e-mail list, or discussion group: *once you have tried it out, you can unsubscribe by sending mail to the same address with no subject and "unsubscribe Webtoday-1" in the body of the message.*

un•sup•port•ed /ˌənsəˈpôrt̲id/ ▸ **adj.** (of a program, language, or device) not having assistance for the user available from a manufacturer or system manager.

un•wired /ˌənˈwī(ə)rd/ ▸ **adj.** disengaged or disconnected from electronic media: *the only way to truly regenerate yourself enough to be truly creative and inventive again is to be unwired at times in the year and to be in the other part of the world.*

un•zip /ˌənˈzip/ ▸ **v.** (**un•zipped, un•zip•ping**) [trans.] decompress (a file) that has previously been compressed.

up /əp/ ▸ **adj.** [predic.] (of a computer system) functioning properly: *the system is now up.*

PHRASES **up and running** (of a computer or other device) in operation; functioning: *the new computer is up and running.*

ORIGIN Old English *up(p)*, *uppe*, of Germanic origin; related to Dutch *op* and German *auf*.

up•date /ˈəpˌdāt/ ▸**n.** a revision to a software program: *download an update.*

▸**v.** [trans.] apply a revision to (a software program): *make sure you regularly update anti-virus software.*

DERIVATIVES **up•dat•a•ble adj.**

up•grade /ˈəpˌgrād/ ▸**n.** an improved or more modern version of software or a piece of computing equipment: *a memory upgrade.*

▸**v.** [trans.] acquire or add an improved or more modern version of (software or a piece of computing equipment): *you should upgrade to version 5.*

DERIVATIVES **up•grad•a•bil•i•ty** (also **up•grade•a•bil•i•ty**) **n.**; **up•grad•a•ble** (also **up•grade•a•ble**) **adj.**

up•load /ˌəpˈlōd/ ▸**v.** [trans.] transfer (data) to a computer from a different computer or another device: *first, upload the photos onto the computer.*

▸**n.** the act or process of transferring data in such a way: *a 5-minute upload.*

UPS ▸**abbr.** uninterruptible power supply.

up•time /ˈəpˌtīm/ ▸**n.** time during which a computer is in operation.

URL ▸**abbr.** uniform (or universal) resource locator, the address of a World Wide Web page.

USB ▸**abbr.** universal serial bus, a connection technology for attaching peripheral devices to a computer which provides fast data exchange.

USB flash drive /ˈyo͞o ˈes ˈbē ˈflæsн ˌdrīv/ ▸**n.** an external flash drive, small enough to carry on a key ring, that can be used with any computer with a USB port.

USB ▸ **n.** Universal Serial Bus, an external peripheral interface standard for communication between a computer and add-on devices such as audio players, joysticks, keyboards, telephones, scanners, and printers.

Use•net /ˈyo͞ozˌnet; ˈyo͞os-/ ▸**n.** a part of the Internet that stores and handles the transmission of messages for newsgroups and discussion groups: *for many, Usenet is the heart of the Internet, because* *it is where the community of its millions of users comes to gather and exchange ideas and views.*

us•er-de•fin•a•ble /ˈyo͞ozər diˌfīnəbəl/ ▸**adj.** having a function or meaning that can be specified and varied by a user.

us•er-friend•ly /ˈyo͞ozər ˈfrendlē/ ▸**adj.** (of software or a computer) easy to use or understand: *the search software is user-friendly.*

DERIVATIVES **us•er-friend•li•ness n.**

us•er-hos•tile /ˈyo͞ozər ˈhästl/ ▸**adj.** (of software or a computer) difficult to use or understand.

us•er in•ter•face /ˈyo͞ozər ˈintərˌfās/ ▸ **n.** the software by which the user and a computer system interact, typically consisting of graphic displays with menus, clickable options, and areas for input: *Macintosh users aren't being ignored here, but they have a much different user interface than either DOS or Unix.*

us•er•name /ˈyo͞ozərˌnām/ ▸**n.** an identification used by a person with access to a computer, network, or Web site.

us•er-o•ri•ent•ed /ˈyo͞ozər ˌôrēˌentid/ ▸**adj.** (of software or a computer) designed with the user's convenience given priority: *a user-oriented application.*

u•til•i•ty /yo͞oˈtilitē/ ▸**n.** (pl. **u•til•i•ties**) a utility program.

ORIGIN late Middle English: from Old French *utilite*, from Latin *utilitas*, from *utilis* 'useful.'

u•til•i•ty pro•gram /yo͞oˈtilitē ˌprōgræm/ ▸**n.** a program for carrying out a routine function.

V

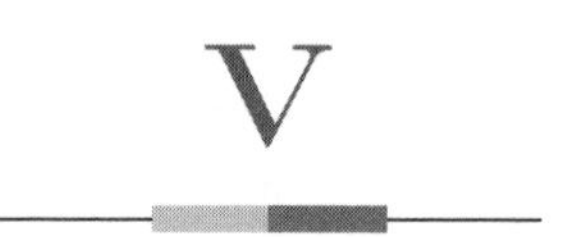

va•por•ware /ˈvāpərˌwe(ə)r/ (Brit. **va•pour•ware**) ▸ **n.** informal software or hardware that has been advertised but is not yet available to buy, either because it is only a concept or because it is still being written or designed.

VAR ▸ **abbr.** value-added reseller, a company that adds extra features to products it has bought before selling them on.

var•i•a•ble /ˈve(ə)rēəbəl/ ▸ **n.** a data item that may take on more than one value during or between programs.

ORIGIN late Middle English: via Old French from Latin *variabilis*, from *variare*.

vCard /ˈvēˌkärd/ ▸ **trademark** virtual business card, an electronic representation of a business card, usually a file attached to an e-mail in place of a signature.

VDT ▸ **abbr.** video (or visual) display terminal.

vec•tor /ˈvektər/ ▸ **n.** [as adjective] denoting a type of graphical representation using straight lines to construct the outlines of objects.

ORIGIN mid 19th cent.: from Latin, literally ‘carrier,’ from *vehere* ‘convey.’

vec•tor proc•es•sor /ˈvektər ˌpräsesər/ ▸ **n.** a processor that is able to process sequences of data with a single instruction.

ver•sion /ˈvərZHən/ ▸ **n.** a particular updated edition of a piece of computer software.

ORIGIN late Middle English: from French, or from medieval Latin *versio(n-)*, from Latin *vertere* ‘to turn.’

ver•sion con•trol /ˈvərZHən kənˌtrōl/ ▸ **n.** the task of keeping a software system that consists of many versions and configurations well organized.

VESA ▸ **abbr.** Video Electronics Standards Association, an organization that defines formats for displays and buses used in computers.

VGA ▸ **abbr.** video graphics array, a standard for defining color display screens for computers.

vid•e•o card /ˈvidēˌō ˌkärd/ ▸ **n.** a printed circuit board controlling output to a display screen.

vid•e•o dis•play ter•min•al /ˈvidēˌō diˌsplā ˌtərmənl/ (also **vis•u•al dis•play ter•mi•nal** /ˈvizHo͞oəl/) (abbr.: **VDT**) ▸ **n.** a computer display, typically a CRT monitor.

vid•e•o game /ˈvidēˌō ˌgām/ ▸ **n.** a game played by electronically manipulating images produced by a computer program on a screen.

vid•e•o•graph•ics /ˌvidēōˈgræfiks/ ▸ **plural n.** visual images produced using computer technology.

■ [treated as sing.] the manipulation of video images using a computer.

vid•e•o pill /ˈvidēō ˌpil/ ▸ **n.** a capsule containing a tiny camera that, when swallowed, transmits photographs of the stomach and intestines to a recording device: *video pills can be used to diagnose ulcers.*

view•er /ˈvyo͞oər/ ▸ **n.** a program that allows a user to view a file but not alter it in any way: *open the PDF file in your normal PDF viewer.*

view•port /ˈvyo͞oˌpôrt/ ▸ **n.** a framed area on a display screen for viewing information.

vir•tu•al /ˈvərCHo͞oəl/ ▸ **adj.** not physically existing, but made by software to appear to do so: *a virtual computer.* See also **VIRTUAL REALITY**.

DERIVATIVES **vir•tu•al•i•ty** /ˌvərCHo͞oˈælit̲ē/ **n.**

vir•tu•al•ize /ˈvərCHo͞oəˌlīz/ ▸ **v.** convert (something) to a computer-generated simulation of reality: [trans.] *traditional universities have begun to virtualize parts of their curricula* | [intrans.] *our method makes it easy to virtualize.*

DERIVATIVES **vir•tu•al•i•za•tion n.**; **vir•tu•a•li•zer n.**

vir•tu•al•ly /ˈvərCHo͞oəlē/ ▸ **adv.** by means of virtual reality techniques: *the program allows you to virtually fly.*

vir•tu•al mem•o•ry /ˈvərCHo͞oəl ˈmem(ə)rē/ (also **vir•tu•al stor•age** /ˈvərCHo͞oəl ˈstôrij/) ▸ **n.** memory that appears to exist as main storage although most of it is supported by data held in secondary storage, transfer between the two being made automatically as required.

vir•tu•al of•fice /ˈvərCHo͞oəl ˈôfis; ˈäfis/ ▸ **n.** the operational domain of any business or organization whose work force includes a significant proportion of workers using technology to perform their work at home.

vir•tu•al pet /ˈvərCHo͞oəl ˈpet/ ▸ **n.** see **CYBERPET.**

vir•tu•al pri•vate net•work /ˈvərCHo͞oəl ˈprīvit ˈnetˌwərk/ (abbr.: **VPN**) ▸ **n.** a method employing encryption to provide secure access to a remote computer over the Internet.

vir•tu•al re•al•i•ty /ˈvərCHo͞oəl rēˈælitē/ ▸ **n.** the computer-generated simulation of a three-dimensional image or environment that can be interacted with in a seemingly real or physical way by a person using special electronic equipment, such as a helmet with a screen inside or gloves with sensors.

vi•rus /ˈvīrəs/ ▸ **n.** (also **com•put•er vi•rus**) a piece of code that is capable of copying itself into other files and typically has a detrimental effect, such as corrupting the system or destroying data.

ORIGIN late Middle English (denoting the venom of a snake): from Latin, literally 'slimy liquid, poison.' The earlier medical sense, superseded by the current use as a result of improved scientific understanding, was 'a substance produced in the body as the result of disease, esp. one that is capable of infecting others with the same disease.'

VLSI ▸ **abbr.** very large-scale integration, the process of integrating hundreds of thousands of components on a single silicon chip.

voice rec•og•ni•tion /ˈvois rekəgˌnisHən/ ▸ **n.** computer analysis of the human voice, especially for the purposes of interpreting words and phrases or identifying an individual voice.

VoIP (also **voice o•ver IP** /ˈvois ˌōvər ˈī ˈpē/) ▸ **n.** voice over Internet protocol, a technology for digital voice communication over the Internet using a standard telephone.

vol•a•tile /ˈvälətl/ ▸ **adj.** (of a computer's memory) retaining data only as long as there is a power supply connected.

DERIVATIVES **vol•a•til•i•ty** /ˌväləˈtilitē/ **n.**

ORIGIN Middle English: from Old French *volatil* or Latin *volatilis*, from *volare* 'to fly.'

vor•tal /ˈvôrtl/ ▸**n.** an Internet site that provides a directory of links to information related to a particular industry.

ORIGIN 1990s: blend of *v(ertical)* (as in *vertical industry*, an industry specializing in a narrow range of goods and services), and *(p)ortal.*

vox•el /ˈväksəl/ ▸**n.** (in computer-based modeling or graphic simulation) each of an array of elements of volume that constitute a notional three-dimensional space, esp. each of an array of discrete elements into which a representation of a three-dimensional object is divided.

ORIGIN 1970s: from the initial letters of *volume* and *element*, with the insertion of *-x-* for ease of pronunciation.

VPN ▸**abbr.** virtual private network.

VR ▸**abbr.** virtual reality.

VRAM ▸**abbr.** video random access memory.

VRML ▸**abbr.** virtual reality modeling language.

MULTIMEDIA AND YOU

While we haven't yet reached the age of true interactive Internet, there are many video, audio and multimedia formats you will encounter that can enrich your computing experience.

Here are several tools which make it work.

The big three are *Windows Media Player*, *Quicktime*, and *RealPlayer*. All three are compatible with Macintosh and Windows, although you may have to download them. Each handles streaming audio and video, and downloadable formats such as MP3 and MPEG. All three programs will automatically download software, if available, when they need it to play a file, and each has plugins which are automatically installed for use by your web browser, so you can play multimedia content directly from a web page.

Be aware that the three programs will battle for control over which program plays which files. All three have preference settings which will allow you to make that program the default player for your chosen formats. However, each also handles a couple of proprietary formats which the others do not, so it's good to have all three.

Besides playing audio and video, the latest versions of Windows Media Player (http://windowsmedia.com/) can help you make audio CDs or import music from CDs to your hard drive. Be aware that if you rip music from your CD collection to certain Windows Media formats, those files might not be playable on other computers. WMP plays files ending in the suffixes .wmv and .wma, among others.

Quicktime (http://www.apple.com/quicktime/) and iTunes (http://www.apple.com/itunes/) together handle audio and video playing, as well as advanced functions of CD-burning and music-importing. iTunes also offers the ability to share song lists over your local network—even between Macs and PCs—to convert music to MP3s, and to interface with an iPod. Quicktime plays files ending in the suffix .mov, among others.

RealOne (http://www.real.com/) is the latest version of RealPlayer, which plays its own proprietary streaming formats, as well as many of the standard formats. It comes in free and pay versions, although you may have to dig for the free version. RealOne plays files ending with the suffixes .rm and .ram, among others.

Sometimes you'll encounter files which won't play with any of the above. There are alternatives:

VideoLAN Client (VLC) is a free program which handles diverse video and audio formats (including DVD discs) for nine different operating systems, including Windows, Macintosh, and Linux. http://www.videolan.org/vlc/

MPlayer is another free program which handles many audio and video formats. It's available for Mac OS X (http://mplayerosx.sourceforge.net/) and Linux (http://www. mplayer-hq.hu/). It can even handle some of the proprietary Quicktime and WindowsMedia formats.

On Windows, WinAmp (http://www.winamp.com/) is the grand-daddy of MP3 players, offering equalizers, play lists, and a variety of different appearances, special effects, and plugins to choose from. There are excellent free versions, and pay versions with more advanced features.

WAIS /wās/ ▸ **abbr.** wide area information service, designed to provide access to information across a computer network.

wait state /ˈwāt ˌstāt/ ▸ **n.** the condition in which computer software or hardware is unable to process further instructions while waiting for some event such as the completion of a data transfer.

wall•pa•per /ˈwôlˌpāpər/ ▸ **n.** an optional background pattern or picture on a computer screen: *desktop wallpaper that you can download to your computer.*

WAN /wæn/ ▸ **abbr.** wide area network.

ORIGIN 1980s: acronym.

war•chalk•ing /ˈwôrˌCHôkiNG/ (also **war chalk•ing**) ▸ **n.** the practice of marking chalk symbols on sidewalks and other outdoor surfaces to indicate the location of unsecured wireless network connections: *savvy IT managers check their buildings' facades for signs of warchalking.*

war di•al•er /ˈwôr ˌdī(ə)lər/ ▸ **n.** a program used to find phone numbers that connect to a modem, often used by someone seeking to access the computers of others without permission.

war•driv•ing /ˈwôrˌdrīviNG/ ▸ **n.** the practice of seeking out and taking advantage of free connection to unsecured wireless networks: *management recently heard about wardriving and is very concerned that corporate information is being intercepted by those who don't need to see it.*

ORIGIN 2003: allegedly coined by Pete Shipley, a San Francisco Bay-area IT consultant, from *war* + *driving*, described by him as "driving around looking for unsecured wireless networks."

wave•ta•ble /ˈwāvˌtābəl/ ▸ **n.** a file or memory device containing data that represents a sound such as a piece of music.

wav file /ˈwāv ˌfīl/ (also **wave file** /ˈwāv ˌfīl/) ▸ **n.** a format for storing audio files that produces CD-quality audio: *save the recording as a wav file on your hard drive.*

wear•a•ble /ˈwe(ə)rəbəl/ (also **wear•a•ble com•put•er** /ˈwe(ə)rəbəl kəmˈpyo͞otər/) ▸ **n.** a computer that is small or portable enough to be worn or carried on one's body: *wearable computers are the height of geek chic.*

web /web/ ▸ **n.** (**the Web**) short for **WORLD WIDE WEB.**

Webcam /ˈwebˌkæm/ (also **Web cam**) ▸ **n.** a video camera attached to a computer and used to transmit video images over the Internet: *she set up a Webcam for the boy to record a message for his father overseas.*

web•cast /ˈwebˌkæst/ (also **Web•cast**) ▸ **n.** a live broadcast on the Internet: *via live audio webcast* | *the news at noon can be seen live via Webcast every weekday.*

DERIVATIVES **web•cast•ing n.**

web-en•a•ble /ˈweb enˌābəl/ ▸ **v.** [trans.] make accessible via or compatible with the World Wide Web: *a project to web-enable legacy accounting systems* | *web-enable your small business.*

DERIVATIVES **web-en•a•bled adjective:** *though each of these three companies is Web-enabled, their online business has yet to overtake their more traditional customers.*

web host•ing /ˈweb ˌhōstiNG/ ▸ **n.** the activity or business of providing storage space and access for Web sites.

web•log /ˈwebˌläg/ ▸ **n.** a Web site on which an individual or group of users produce an ongoing narrative: *he's been writing a weblog and keeping an online photo diary.*

DERIVATIVES **web•log•ger n.**

ORIGIN 1990s: from *web* in the sense 'World Wide Web' and *log* in the sense 'regular record of incidents'.

Web•mail /ˈwebˌmāl/ (also **web•mail**) ▸ **n.** e-mail available for use online and stored in the Internet server mailbox, and that is not downloaded to an e-mail program or used offline.

Web•mas•ter /ˈwebˌmæstər/ (also **web•mas•ter**) ▸ **n.** a person who designs and develops Web sites.

Web page /ˈweb ˌpāj/ (also **web page**) ▸ **n.** a document connected to the World Wide Web and viewable by anyone with an Internet connection and a browser.

Web site /ˈweb ˌsīt/ (also **web site** or **web•site** /ˈwebˌsīt/) ▸ **n.** a location connected to the Internet that maintains one or more pages on the World Wide Web.

well-be•haved /ˈwel biˈhāvd/ ▸ **adj.** (of a computer program) communicating with hardware via standard operating system calls rather than directly and therefore able to be used on different machines.

what-you-see-is-what-you-get /ˈ(h)wət yo͞o ˈsē iz ˈ(h)wət yo͞o ˈget/ ▸ **adj.** see **WYSIWYG**.

white•board /ˈ(h)wītˌbôrd/ ▸ **n.** an area common to several users or applications, where they can exchange information, often as handwriting or graphics.

white•list•ing /ˈ(h)wītˌlistiNG/ ▸ **n.** the use of antispam filtering software to allow only specified e-mail addresses to get through: *whitelisting sometimes backfires because it filters e-mail from people or companies you might be interested in.*

ORIGIN on the pattern of *blacklisting.*

wide ar•e•a net•work /ˈwīd ˈe(ə)rēə ˈnetˌwərk/ (abbr.: **WAN**) ▸ **n.** a computer network in which the computers connected may be far apart, generally having a radius of half a mile or more.

wide•screen /ˈwīdˌskrēn/ (also **wide-screen**) ▸ **adj.** [attrib.] (of a computer monitor) designed with or for a screen presenting a wide field of vision in relation to its height: *a widescreen notebook.*

▸ **n.** (**wide•screen**) a monitor screen with a wide field of vision in relation to its height.

■ a film format presenting a wide field of vision in relation to height.

widg•et /ˈwijit/ ▸ **n.** informal a component of a user interface that operates in a particular way.

ORIGIN 1930s : perhaps an alteration of *gadget.*

Wi-Fi /ˈwī ˈfī/ ▸ **trademark** Wireless Fidelity, a group of technical standards enabling the transmission of data over wireless networks. *coffeehouses are installing Wi-Fi to make it easier for customers to do business while sipping cappuccino.* Also see **IEEE 802.11**.

wild card /ˈwīld ˌkärd/ ▸ **n.** a character that will match any character or sequence of characters in a search.

WIMP /wimp/ ▸ **n.** [often as adjective] a set of software features and hardware devices (such as windows, icons, mice, and pull-down menus) designed to simplify or demystify computing operations for the user.

ORIGIN 1980s: acronym.

win•dow /ˈwindō/ ▸ **n.** a framed area on a display screen for viewing information: *right-click on the link and it will open in a new window.*

ORIGIN Middle English: from Old Norse *vindauga*, from *vindr* 'wind' + *auga* 'eye.'

win•dowed /ˈwindōd/ ▸ **adj.** having or using framed areas on a display screen for viewing information.

win•dow•ing /ˈwindō-iNG/ ▸ **n.** the use of windows for the simultaneous display of more than one item on a screen.

Win•dows /ˈwindōz/ ▸ **trademark** [treated as sing.] a computer operating system with a graphical user interface.

wipe /wīp/ ▸ **v.** [trans.] erase (data) from a magnetic medium.

ORIGIN Old English *wīpian*, of Germanic origin.

wired /wī(ə)rd/ ▸ **adj.** informal making use of computers and information technology to transfer or receive information, esp. by means of the Internet.

■ denoting the availability of Internet access: *the museum has a wired cafe-bar.*

wire fraud /ˈwī(ə)r ˌfrôd/ ▸ **n.** financial fraud involving the use of telecommunications or information technology.

wire•less /ˈwī(ə)rlis/ ▸ **adj.** denoting communication or the ability to communicate over a system using radio waves, without a physical connection by wire or cable: *wireless Internet access* | *a wireless notebook or PDA.*

wire•less hot•spot /ˈwī(ə)rlis ˈhätˌspät/ ▸ **n.** an area with a usable signal to allow wireless connection to the Internet or some other computer network: *a wireless hotspot where people can surf the Internet and check their e-mail without even plugging in their laptops*

wire•less•ly /ˈwī(ə)rlislē/ ▸ **adv.** without a wire connection; using a wireless technology: *a patented FM technology which broadcasts music wirelessly from a small transmitter to satellite speakers.*

wire•less net•work /ˈwī(ə)rlis ˈnet,wərk/ ▸ **n.** a network in which computers do not connect by wires or cable, but use radio waves to connect and exchange information: *I recently set up a wireless network at home.*

wiz•ard /ˈwizərd/ ▸ **n.** a help feature of a software package that automates complex tasks by asking the user a series of easy-to-answer questions: *the included installation and setup wizard.*

ORIGIN late Middle English: from *wise* + *-ard.*

WLAN /ˈdəbəlyo͞o ˌlæn/ ▸ **abbr.** wireless local area network.

word /wərd/ ▸ **n.** a basic unit of data in a computer, typically 16 or 32 bits long.

ORIGIN Old English, of Germanic origin; related to Dutch *woord* and German *Wort*, from an Indo-European root shared by Latin *verbum* 'word.'

word length /ˈwərd ˌleNG(k)TH/ ▸ **n.** the number of bits in a word.

word proc•ess•ing /ˈwərd ˌpräsesiNG/ ▸ **n.** the production and manipulation of text on a computer, using word processing software.

DERIVATIVES **word-proc•ess v.**

word proc•es•sor /ˈwərd ˌpräsesər/ ▸ **n.** a program for storing, manipulating, and formatting text on a computer: *the software suite does include a word processor.*

word wrap /ˈwərd ˌræp/ ▸ **n.** see **WRAP.**

work•a•like /ˈwərkəˌlīk/ ▸ **n.** software that is identical in function to another software package: *an Office workalike productivity suite.* ■ a computer that is able to use the software of another specified machine without special modification.

work•book /ˈwərkˌbo͝ok/ ▸ **n.** a single file containing several different types of related information as separate worksheets.

work•group /ˈwərkˌgro͞op/ ▸ **n.** a group who share data via a local network.

work•ing mem•o•ry /ˈwərkiNG ˈmem(ə)rē/ ▸ **n.** an area of high-speed memory used to store programs or data currently in use.

work•ing stor•age /ˈwərkiNG ˈstôrij/ ▸ **n.** a part of a computer's memory that is used by a program for the storage of intermediate results or other temporary items.

work•sheet /ˈwərkˌSHēt/ ▸ **n.** a data file created and used by a spreadsheet program, which takes the form of a matrix of cells when displayed.

work•sta•tion /ˈwərkˌstāsHən/ ▸ **n. 1** a desk with a computer. **2** a general-purpose computer with a higher performance level than a personal computer.

World Wide Web /ˈwərld ˈwīd ˈweb/ a widely used information system on the Internet that provides facilities for documents to be connected to other documents by hypertext links, enabling the user to search for information by moving from one document to another.

WORM /wərm/ ▸ **abbr.** write-once read-many, denoting a type of computer memory device.

worm /wərm/ ▸ **n.** a self-replicating program able to propagate itself across a network, typically having a detrimental effect: *the worm tries to get user information by sending a fake eBay e-mail that says billing information is out of date.*

DERIVATIVES **worm•like** /ˈwərmˌlīk/ **adj.**

ORIGIN Old English *wyrm* (noun), of Germanic origin; related to Latin *vermis* 'worm' and Greek *rhomox* 'woodworm.'

WP ▸ **abbr.** word processing or word processor.

wrap /ræp/ ▸ **v.** (**wrapped**, **wrap•ping**) [trans.] cause (a word or unit of text) to be carried over to a new line automatically as the margin is reached, or to fit around embedded features such as pictures.

■ [intrans.] (of a word or unit or text) be carried over in such a way.

ORIGIN Middle English: of unknown origin.

wrap•per ap•pli•ca•tion /ˈræpər æpliˌkāsHən/ ▸ **n.** a computer program that works only with another fully developed program, which it enhances in some way: *we have created a viewer which is a simple wrapper application for the underlying multimedia system.*

write /rīt/ ▸ **v.** (past **wrote** /rōt/; past part. **writ•ten** /ˈritn/) [trans.] enter (data) into a specified storage medium or location.

DERIVATIVES **writ•a•ble adj.**

ORIGIN Old English *wrītan* 'score, form (letters) by carving, write,' of Germanic origin; related to German *reissen* 'sketch, drag.'

write-once /ˈrīt ˈwəns/ ▸ **adj.** denoting a memory or storage device, typically an optical one, on which data, once written, cannot be modified.

write-pro•tect /ˈrīt prəˌtekt/ ▸ **v.** [trans.] protect (a disk) from accidental writing or erasure.

WWW ▸ **abbr.** World Wide Web.

WXGA ▸ **abbr.** wide extended graphics array, a standard for high-resolution widescreen monitors.

WYSIWYG /ˈwizēˌwig/ (also **wysiwyg**) ▸ **adj.** denoting the representation of text on screen in a form exactly corresponding to its appearance on a printout.

ORIGIN 1980s: acronym from *what you see is what you get.*

TEN GREAT COMPUTER GAMES YOU SHOULD (OR MAYBE SHOULDN'T) KNOW

Computers aren't just for writing letters, calculating how much money you've lost in the stock market, or e-mailing the entire family about the cancer-stricken boy who is collecting business cards (or is it get-well cards?). No, computers also have a more useful purpose—gaming. Here are ten games that you can waste many hours of your life playing.

Tetris (http://www.tetris.com)—Polygons fall from the sky. Make rows of bricks out of the polygons. Repeat. Simple. . .until the polygons are dropping so fast you have less than second to find a place for the current piece and the incomplete rows start building up. . .game over. So you play again. . .you're addicted. Simple enough for a three year old to play, yet so baffling that the kid will probably outscore you. University students have turned high-rise buildings into giant Tetris games, the record holder being a 15 story academic tower in the Netherlands.

Solitaire (probably on your computer right now)—Along with Minesweeper and Hearts, this is how we wasted time at work in the days before we had Internet access at our desks and could check scores on ESPN's Web site all day.

Civilization I/II/III (http://www.civ3.com/)—A sandbox for your Napoleon complex. Start from the Iron Age and lead your people into the nuclear age by investing in scientific research and exploring new terrain. Make friends with other civilizations, or roll your tanks right up to their doorsteps.

Deus Ex (http://www.deusex.com/)—Like other first-person shooter games (such as Doom and Quake), this game is gory and not for kids. Unlike Doom and Quake, you don't have to shoot everything that moves to win. Stealth, non-lethal methods of neutralizing the enemy, and an open mind will get you through this game of near-future intrigue.

Everquest (http://everquest.station.sony.com/)—The most well known MMORPG (Massively Multiplayer Online Role Playing Game). Dungeons and Dragons players no longer have to leave their house and interact with real people. This is probably the only game on this list given as a reason for divorce.

The Sims (http://thesims.ea.com/)—Not happy with your life? Then create a new one. Design a person, build them a house, get them a better job, find them a spouse, and finally put that swimming pool in the backyard you've always wanted. The Sims was the first PC game that sold big among women who weren't part of the "gamer" subculture.

Zork (http://www.csd.uwo.ca/Infocom/)—The radio of the gaming world, this is a text-only game from the days when gameplayers walked uphill in the snow both ways to play computer games. Despite the lack of graphics (or because of it), it's still an imaginative, challenging, humorous, literate game of dungeon exploration and puzzle solving. (Note: Zork is such an old game you'll need a special program called a Z-Machine interpreter to play it. You can download one for your particular system from this page: http://www.csd.uwo.ca/Infocom/interp. html).

Bejeweled (online; Mac, PC, and Palm versions available from http://www.popcap.com)—Similar to Tetris and likewise simple to learn, hard to master. It can be played on a multitude of computing systems, including handheld devices. Are you certain that your co-worker is really taking notes on her Palm Pilot during those endless meetings?

Microsoft Flight Simulator (http://www.microsoft.com/games/flightsimulator/)—Learning to fly a plane on a computer? Fun. Discovering that a real cockpit doesn't have a mouse or keyboard to control the plane? Not so fun.

Simcity (http://simcity.ea.com)—It should have been a dud of a game. Urban planning doesn't sound like a subject to inspire all-night play. Yet the many versions of Simcity have sold millions of copies, inspiring future civil engineers. Either we all want to build Utopia, or we just want to build a Utopia so we can destroy it with tornados, fires, and UFO attacks.

Sound like fun? Then check out the URLs and download or buy the games. You may want to invest in a comfy computer chair and a Colombian coffee plantation, though, because computer games are extremely addictive. Don't try playing them all at once, and remember to eat, sleep, and get away from the computer once in a while. Virtual worlds are great, but until the day you can download a cheeseburger and a side order of onion rings, the real world is better.

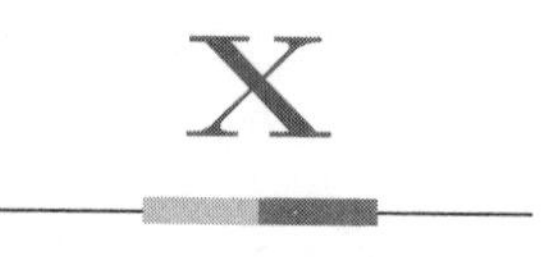

XGA ▸ abbr. extended graphics array, a standard for high-resolution monitors and screens.

XHTML ▸ abbr. extensible hypertext markup language, an HTML system for tagging text files to achieve font, color, graphic, and hyperlink effects on World Wide Web pages, incorporating user-defined elements.

XML ▸ abbr. Extensible Markup Language; a compliant version of SGML that is designed to make a wide variety of documents compatible with the Internet and electronic data exchange: [often as modifier] *unlike the existing HTML format, the XML-enhanced format will support features such as pivot tables in spreadsheets, and revision marks in word documents.*

XMS ▸ abbr. extended memory specification, a system for increasing the amount of memory available to a personal computer.

zip /zip/ ▸**n.** a format for compressing a file.

▸**v.** (**zipped**, **zip•ping**) compress (a file) so that it takes less space in storage or as an attachment to an e-mail: *she zipped the documents and sent them by e-mail.*

ORIGIN mid 19th cent.: imitative.

Zip drive /ˈzip ˌdrīv/ ▸**trademark** a disk drive that stores data on high-capacity removable magnetic disks, often used for data backup.

zip file /ˈzip ˌfīl/ ▸**n.** a compressed file having the extension .zip: *the e-mail includes a zip file.*

■ a number of individual files that have been compressed into a single file.

DO YOU NEED A PDA?

You've seen them and they look seductive, those tiny electronic devices that are half appointment book and half *Star Trek* tricorder. But do you really need one?

Do you already have an organizer that works? If you're committed to your paper planner or to a calendar function on your desktop computer, it may be more trouble than it's worth to switch to a PDA.

Do you travel often? Travelers often find PDAs attractive for two reasons: their small size and the automatic backup of data to a computer at home or work. Losing a paper planner while traveling can be traumatic. Losing a PDA is a little less so, since at least you know that you haven't lost the information, just the information container.

Are you happy with gadgets? If your VCR is unprogrammed and you regularly swear at your desktop computer, a PDA might be more than you want to wrestle with. However, most PDAs have good interfaces and can be learned quickly, even by the most tech-averse of users.

What kind of information do you need to access? If you only need a basic calendar, names and phone numbers, and a calculator, there are smaller, less expensive electronic devices that may meet your needs at any office supply store. Check them out before springing for a larger PDA.

Will your desktop computer support a PDA? Many older computers may not have the USB (universal serial bus) connectors that modern PDAs require. Adapters are available but are an extra expense and may add to your setup and troubleshooting time. You don't want to have to upgrade your entire system just to use a PDA.

Will your desktop software support a PDA? One of the main advantages to using a PDA is being able to synchronize data from your desktop applications to the PDA. If your company mandates a particular e-mail, meeting coordination, or calendar software, make sure that it will 'sync' with your PDA. Otherwise, you'll spend a lot of time manually entering information, or wishing you had.

Do you already carry another electronic device? If you already carry a cell phone or digital camera for work, you might want to look into multipurpose PDAs. However, only make the switch if the multipurpose device fulfills all your requirements for both devices.

Do you lose things? If you go through a dozen pens a week, two pairs of reading glasses a month, and replace your keys twice a year, you may not be the best candidate for a PDA.

Do you need specialized information regularly? There are many add-on software titles for PDAs that deliver everything from prescription drug interaction information to the location of the closest Chinese restaurant to stock portfolio management. If having such information on-the-go would be useful to you, consider a PDA.

Will it improve your life? If a PDA is just one more thing to keep track of and worry about, stick to your paper planner or current system. However, if you think you would use a PDA to the fullest—including reading e-books or playing games or keeping track of your personal finances—there are many web pages that will guide you through buying the one that's right for you.

50 WEB SITES YOU SHOULD KNOW

Ain't It Cool News
http://www.aicn.com/
Rabid fans gossip, preview, and review movies. The Internet in microcosm: unruly, but ultimately rewarding.

Arts & Letters Daily
http://www.artsandlettersdaily.com/
A continuously revised collection of links to thoughtful articles on thinking subjects.

AskOxford
http://www.askoxford.com/
Oxford University Press offers the best answers to common language-related questions, free words-of-the-day e-mails, and tools for better writing.

B3ta
http://www.b3ta.com/
British site with goofy images, hilarious image-editing contests, and endless punter commentary in the forums.

Babel Fish
http://babelfish.altavista.com/
Translate text and web sites to and from dozens of languages. Not perfect, but as good as it gets for free.

Bartleby
http://www.bartleby.com/
As an impressive collection of scholarly reference works and classic literature, and a site you can quote with certainty and authority.

Blogdex
http://blogdex.net/
Keeps track of the most-linked new content on the Internet: memes, news, jokes, videos, whatever.

BoingBoing
http://www.boingboing.net/
A counter-culture group blog, presenting novel and interesting items from the front edge of pop culture.

Bookfinder
http://www.bookfinder.com/
A search engine for book sites which even searches other book search engines.

CIA World Factbook
http://www.cia.gov/cia/publications/factbook/
Look what the spooks are up to! They've been collecting an accurate almanac about every country around the world, including details on political systems, weather, geography, ethnic groups, maps, and current problems.

Craigslist
http://www.craigslist.org/
Free to visitors, this online classifieds site is the best on the Internet for housing, jobs, dating and other categories. Its self-policing community offers the best ads with the least amount of false leads, fakes, and get-rich-quick schemes, although there's no telling what kind of weirdos will turn up in the personals.

Daypop
http://www.daypop.com/
A current events search engine for news and commentary, including thousands of weblogs, those personal sites where people write what's on their minds.

EatonWeb Portal
http://portal.eatonweb.com/
A vast directory of weblogs sorted by country, language, category, and name.

Family Search
http://www.familysearch.org/
Research your roots in this free archive of genealogical records.

Fark
http://www.fark.com/
An endless supply of wacky news, goofy links, audio- and photo-editing challenges, and a bit of mild adult content.

FedStats
http://www.fedstats.gov/
Statistics from more than 100 federal agencies, including easy-to-understand breakdowns of census records, and links to the original sources.

Feedster
http://www.feedster.com/
A collator of feeds from weblogs, and a great place to search when you want to see what everyone's saying about anything.

Gawker
http://www.gawker.com/
Gossip about the rich, the famous, and the wannabes. Absolutely a waste of time, but so much fun.

Go Ask Alice
http://www.goaskalice.columbia.edu/
Very frank and informed advice on health—including sexual health—from Columbia University. Particularly suited for delicate or "just wondering" questions.

Google News
http://news.google.com/
An automated news collator, collecting news stories from more than 4500 news sites around the world, for numerous languages and countries.

Google
http://www.google.com/
The big daddy of search engines: fast and accurate.

HyperHistory
http://www.hyperhistory.com/
The site is ugly, but so very useful, offering a time-line littered with key names and events, each linking to a nugget of information.

IMDB
http://www.imdb.com/
The Internet Movie Database is a vast resource of film-related information, listing practically every movie ever made and anyone who starred in them.

InfoGrid
http://www.infogrid.com/
Portals abound on the Internet, but this site concentrates on site quality and comprehensiveness.

Internet Traffic Report
http://www.internettrafficreport.com/
If your Internet connection is slow, it could be the result of an interruption thousands of miles away. This site constantly mon-

itors the Internet, looking for failures and delays, and reports with easy-to-understand charts and graphs.

Issue Focus Reports

http://usinfo.state.gov/products/medreac.htm Produced by the State Department, these reports summarize foreign media reaction to current world events.

Library of Congress

http://www.loc.gov/

The American Memory section of the LOC is a vast repository of video, images, sounds, and manuscripts, most available in detail for download or inspection.

Metafilter

http://www.metafilter.com/

Called MeFi to those in the know, members of this group blog post and discuss anything which interests them. A sub-site, Ask Metafilter (http://ask.metafilter.com/), lets any user pose questions to which other MeFites help find answers.

MetaSpy

http://www.metaspy.com/

People are freaks, and they can't spell. MetaSpy proves it: see what people are searching for on the Internet. Choose a clean or adult version.

MIT OpenCourseWare

http://ocw.mit.edu/

Free online classes from one of the best universities on the planet.

NewsLink

http://newslink.org/

Still don't have enough news? Find local media for cities around the world, large and small, in this comprehensive link set sorted by most-linked, type, and location.

Perry-Castañeda Library Map Collection
http://www.lib.utexas.edu/maps/
Links to thousands of current and historical maps, political and geographical, for everywhere.

Planetary Photojournal
http://photojournal.jpl.nasa.gov/
NASA has thousands of exquisite images from space, showing our solar system, the galaxy, and the universe beyond.

Project Gutenberg
http://promo.net/pg/
Download and read at your leisure the text of thousands of books now in the public domain.

Public Radio Fan
http://www.publicradiofan.com/
This fellow has put together a catalog of online broadcasters, their schedules, and links to the streaming audio. See what's on now, by station, language, or type of programming.

RefDesk
http://www.refdesk.com/
A portal, they call it, but in reality it's a link festival, pointing to great reference sites all over the Internet. You could spend days here.

Romanesko's Letters at Poynter Forums
http://poynter.org/forum/?id=letters
If you ever thought the media was a monolith of thought and action, the discord and critiques posted here will give you hope. Journalists are their own best critics, and are not about to let their colleagues get away with *anything*.

Rotten Tomatoes
http://www.rottentomatoes.com/
Never see a stinker in the movie theatre again! Dozens of movie reviews from around the country collected, quoted, and counted, giving each movie an overall rating.

SERVEnet
http://www.servenet.org/
Thousands of opportunities to volunteer your time, mind, and muscles to charities and worthy causes around the country.

Slashdot
http://www.slashdot.org/
Computer and tech-oriented news links, interviews, and reviews about cutting-edge technology, theory, politics, followed by a lot of free-for-all goofing off in the discussion forums.

Snopes
http://www.snopes.com/
The best source for debunking hoaxes, rumors, conspiracy theories, and even bogus computer virus warnings.

The Mirror Project
http://www.mirrorproject.com/
Inexplicably fascinating: thousands of pictures of people in reflective surfaces.

The Obscure Store & Reading Room
http://www.obscurestore.com/
Odd news. Very odd news. Weird, too. Also startling, appalling, laughable, and asinine.

Today's Front Pages
http://www.newseum.org/todaysfrontpages/
Images of newspapers around the world permit you to get a sense of what issues other countries consider important.

UNESCO Photobank

http://upo.unesco.org/photobank.asp More than 10,000 images from around the world have been collected here, from the fantastic to the mundane.

Versiontracker

http://www.versiontracker.com/

Keep track of the latest software releases and updates for Windows, Macintosh, and Palm.

Weather Underground

http://www.wunderground.com/

See the weather and the forecasts for just about anywhere on the planet.

Web Archive

http://www.archive.org/

Only just getting on the Internet? You've got catching up to do! See the Net as it used to be in this awesome archive of web data which has elsewhere disappeared.

Worth1000

http://www.worth1000.com/

Nothing but image-editing contests, producing oodles of lifelike but unreal results based around themes.

Yahoo Games

http://games.yahoo.com/

More online fun than you can shake a joystick at, from whist and bridge to fantasy sports leagues.

A QUICK GUIDE TO WRITING ONLINE ENGLISH

What is Online English? Online English is simply the English that people use online. That's it. That's all. Unfortunately, "Online English" when described that way, seems to bring with it an air of either trepidation or disdain. Trepidation, in that many people believe that they will not be able to understand what "these kids today" are writing online, or disdain, in that they can't imagine why anyone would care what "these kids today" are writing online. The mental image is of a chatroom screen filled with TLAs (three-letter acronyms), emoticons, and enough exclamation points to outfit an entire shelf of melodramatic novels.

The emoticons and acronyms are but a small part of Online English. The secret of Online English is this: it's very close to a language you already know, and know well: Informal English.

The confusion comes about because, offline, you speak Informal English much more than you write it. Sure, you might dash off a quick postcard to a friend, or leave a sticky note on a co-worker's chair, but for the most part, when you are using Informal English, you're speaking it. Online, of course, you write much more than you talk—it's just that your online writing is (or should be, for the most part) much like conversation.

The different kinds of Online English can be described much like the different kinds of conversation. At the most formal,

Online English can be like the conversation at a professional meeting, with prepared remarks and considered dialogue. Many blogs and academic discussion groups have this tone. The point of these online discussions is to get ideas across clearly and succinctly, with a certain amount of style—not stiltedly or slangily, but certainly with the end goal of clarity and mutual understanding. Many blogs, if printed out and divorced from their web associations, would read just as well as many newspaper op-ed columns. No emoticons here! Think of this level as being the same as an informal business memo.

The next level of Online English is e-mail. The only real rule for Online English in e-mail is that you should be careful to be more explicit and more considerate than you would be in a similar conversation. E-mail is a broad category, and possibly the one that most people are familiar with. Many people think of e-mail as just like talking in the halls at work, only without any boring talk about the weather. An "I need this by five today," in the hallway is usually accompanied by a smile or a shrug—some nonverbal acknowledgment of the request. An "I need this by five today" in e-mail, without an introductory "Dear Stephen, I know this is short notice" and coda of "Thanks! I really appreciate it!" is curt and unfriendly and unlikely to get you any helpful response, much less by five that day. It's helpful to remember that most e-mail is just a letter without a stamp, and that writing an e-mail has nearly the same rules as writing a letter. When you write a formal e-mail (complaining, petitioning, proposing, or introducing), you use nearly all the same parts that you would in a formal letter—a formal salutation, a formal closing, a signature that includes your personal information, including a snail-mail address and phone number. A friendly e-mail is just like a friendly letter, only even more friendly: a "Dear Peter" at the top, a "Yours" or "Best" or "Take care—" at the bottom, and of course standard capitalization and punctuation. Make sure to write in paragraphs! Onscreen, especially, a large block of text with no breaks is offputting. Feel free to slip in an emoticon here and there, but stick to the standard smiley face. (You don't ever need elaborate emoticons of the Pope wearing a propeller beanie, by the way, no matter how cute the idea is.) An even more informal e-mail is just like a quick note: "Here's the file you need! Let me know if you have any questions." (Think of a memo-pad note left on someone's desk or chair.) A line above a forwarded e-mail, or a quick response to a short e-mail, can be simpler still: "Thanks!" "see you soon." "this is great. . ." (Think sticky note. Capitalization and complete punctuation favored, but not mandatory) And I'm sure you know this already, but unless you want to come across as screaming in incoherent rage, NEVER SEND AN E-MAIL IN ALL CAPS.

Bulletin boards and discussion groups are more informal yet. The level of informality varies from site to site. More general groups will be the most informal. The narrower the interest, the closer to standard English the postings will be, as a general rule. Postings on a discussion group devoted to the works of Jane Austen will probably be more formal than the writing on a web site devoted to collecting rubber chickens, although there are always exceptions. Always read some postings before jumping in, and look for (and read) any kind of frequently asked questions (FAQ) list or guide for new members, too. Think of this as like attending an informal meeting—maybe a neighborhood get-together or a club. People will interrupt each other, tell jokes, swap stories, and, in general, behave casually.

Instant messaging (IM) and chat rooms are perhaps the most informal settings for Online English. These are real-time, and thus are the most like phone conversations—phone conversations with your best friend, the one who finishes all your sentences. Speed is important, which is why you see all those acronyms. . .people are trying to keep the flow going. False starts abound; capitalization is a nicety, at best, and exclamation points come in handy. Of course, if you are IM-ing with your boss, you might not want to type "OMG! [oh my god!] that meeting took 4eva!" but if you're IM-ing your teenager, let the acronyms fly. (A short list of the most common acronyms appears at the end of this essay.)

But what about "kewl" ('cool') and "1337" ('elite' a now-jokey term often used by 'elite' computer gamers) and all those other words you don't know (and your kids do)? Think about it: your kids know a lot of words that you don't—teen slang is one of the few joys of being a teen, after all. It's just that some of their slang is used mainly online, instead of on the soccer field or in notes passed in class. (Didn't you ever write "BFF" 'best friends forever' or at least intercept a note signed that way, back when you were a teen?) A lot is used to annoy people in general (and you in particular, it must be said), to signal their membership in their peer group, and some just for the pure fun of it. Even those of legal drinking age and well past use some of the more common online slang and jargon, for any or all of the reasons above. The nicest part, though, of this online jargon is that there are just as many people as confused about it as you are, and more than a few helpful souls who are happy to explain. A quick Google of any term you're unsure of will probably bring up a dozen or more pages that explain it.

Possibly the most important thing to remember about Online English is that you shouldn't feel that you have to make your online writing style less formal—no one will flame ('criticize in an angry way') you for not using acronyms or emoticons. You may sound stilted, but a little stilted is better than coming across as unsure of yourself and uncomfortable. It's also advisable to give your online correspondents the benefit of the doubt. Is that writer using all lower case because she's lazy, or because she's disabled and has trouble using a keyboard? Is "Borg2K" using incomprehensible acronyms to make you feel excluded or has he been using them for so long he didn't even consider that someone wouldn't know them? When in doubt, reread, repost or reframe your question, or ask for clarification. Most of all, enjoy the opportunity to write quickly and informally, to use fun jargon (or not, as you prefer) and to communicate both with those who share your interests and those whose worlds are far removed from your own!

Text Messaging Abbreviations

@	at
ADN	any day now
AFAIK	as far as I know
ATB	all the best
B	be
BBL	be back late(r)
BCNU	be seeing you
BFN	bye for now
B4	before
BRB	be right back
BTW	by the way
BWD	backward
C	see
CU	see you
CUL8R	see you later
F2F	face to face
FWD	forward
FWIW	for what it's worth
FYI	for your information
GAL	get a life
GR8	great
HAND	have a nice day
H8	hate
HTH	hope this helps
IC	I see
IMHO	in my humble opinion (often IMNSHO 'in my not-so-humble opinion')
IOW	in other words
JIC	just in case
JK	just kidding
KIT	keep in touch
KWIM	know what I mean?
L8	late
L8R	later
LOL	laughing out loud
MYOB	mind your own business
NE	any
NE1	anyone
NO1	no one
OIC	oh, I see
OMG	oh my God!
OTOH	on the other hand
PLS	please
POS	parent over shoulder
PPL	people
R	are
ROFL	rolling on the floor laughing
RU	are you
RUOK	are you okay?
SIT	stay in touch
SOM1	someone
THKQ	thank you
TTYL	talk to you later
TX	thanks
U	you
UR	you are
WAN2	want to

W/	with	1	one
WKND	weekend	2	to, too
WU	what's up?	2DAY	today
X	kiss	2MORO	tomorrow
XLNT	excellent	2NITE	tonight
XOXOX	hugs and kisses	4	for
YR	your		

Basic Emoticons

:-) happy
:-(sad
;-) winking
:-P sticking tongue out
:-D laughing
:-o surprised